A Brighter Tomorrow

A Brighter Tomorrow

Fulfilling the Promise of Nuclear Energy

Senator Pete V. Domenici

With
Blythe J. Lyons and
Julian J. Steyn

ROWMAN & LITTLEFIELD PUBLISHERS, INC.
Lanham • Boulder • New York • Toronto • Plymouth, UK

ROWMAN & LITTLEFIELD PUBLISHERS, INC.

Published in the United States of America
by Rowman & Littlefield Publishers, Inc.
A wholly owned subsidiary of The Rowman & Littlefield Publishing Group, Inc.
4501 Forbes Boulevard, Suite 200, Lanham, Maryland 20706
www.rowmanlittlefield.com

Estover Road
Plymouth PL6 7PY
United Kingdom

Distributed by National Book Network

British Library Cataloguing in Publication Information Available

The hardback edition of this book was previously cataloged by the Library of Congress as follows:
Domenici, Pete.
    A brighter tomorrow : fulfilling the promise of nuclear energy / Pete V.
Domenici.
        p. cm.
    Includes bibliographical references and index.
    ISBN-13: 978-0-7425-4188-7 (cloth : alk. paper)
    ISBN-10: 0-7425-4188-6 (cloth : alk. paper)
    ISBN-13: 978-0-7425-4189-4 (pbk. : alk. paper)
    ISBN-10: 0-7425-4189-4 (pbk. : alk. paper)
    1. Nuclear engineering—Popular works. 2. Nuclear power plants—Popular
works. I. Title.
    TK9146.D66 2004
    333.792'4—dc22   2004010521

Printed in the United States of America

♾™ The paper used in this publication meets the minimum requirements of American National
Standard for Information Sciences—Permanence of Paper for Printed Library Materials,
ANSI/NISO Z39.48-1992.

Contents

Tables and Figures

Tables

Figures

Foreword
Senator Sam Nunn

In 1991, when it was clear that the Soviet Union was unraveling, Senator Richard Lugar and I introduced legislation in the U.S. Senate that would help Moscow secure and destroy their nuclear weapons and keep them out of the hands of rogue states or terrorists. It was a novel idea, to be sure, and the initial reaction by a number of senators was almost contemptuous, with some opponents declaring it "aid to the Soviet military."

Changing that mind-set was possible only through the decisive interventions of highly respected members of the Senate—particularly Senator Pete Domenici. The senator from New Mexico, during a crucial point in the debate, made his way to the floor and asked to speak.

He reminded his colleagues of the billions of dollars we had spent trying to check the spread of communism, and how we were obliged to offer assistance to these emerging democracies in their chaotic quest for freedom. If we did not, he asked his colleagues: "Who knows where those nuclear weapons are going to be, and who is going to control them?"

Pete Domenici's appeal made a dramatic impact—both because of the much-admired moral dimension of his personality and the authoritative power of his arguments. Pete is the kind of colleague who makes you want to go back and check your facts if he differs with you on an issue.

Senator Domenici's timely intervention and cosponsorship on the Senate floor was crucial in the passage of the Nunn–Lugar legislation. Pete was the leader in creating from that legislation both our lab-to-lab program and our materials protection program, which are the mainstays of our current security relationship with the former Soviet Union. He also did more than anyone in the Congress in launching the highly enriched uranium purchase program, which is making progress in reducing the amount of bomb-making materials in Russia.

In 1996, Pete joined Dick Lugar and me in coauthoring and passing the "Defense against Weapons of Mass Destruction Act"—America's first concerted effort to train local and state law enforcement officials to respond in the event of a catastrophic terrorist attack. This bill—known as the Nunn–Lugar–Domenici legislation—was the genesis of America's homeland defense efforts, and provided the foundation for government action since September 11.

Pete has been a vital and indispensable player in the post–Cold War effort to protect the American people. That he would take time to write a book on nuclear power is a valuable extension of his public service. Nuclear energy is an important energy source here in the United States and around the world. Yet, the future of nuclear power is not just a technological question; it's a political one. Citizens of the world will always link the safety of nuclear power with the prospect of terrorists acquiring nuclear materials and exploding a nuclear bomb or a dirty bomb. Consequently, there is a direct connection between protecting materials at the source to keep them out of terrorist hands, and promoting nuclear power around the globe.

Pete makes it clear in the pages that follow that nuclear materials and weapons expertise must be kept out of the hands of rogue nations and terrorist groups. We need a system of global nuclear materials management in which all nuclear materials are safe, secure, and accounted for from "cradle to grave," with sufficient transparency to assure the world that this is the case. Nuclear material security, nonproliferation, and safe nuclear power plants and waste disposal procedures all form a chain that must be unbreakable if the world is to realize the full potential of nuclear power in cleaning our air and closing the gap between the world's haves and have-nots. Pete Domenici has worked at the intersection of all of these crucial issues.

Pete strongly believes, as he writes, "that nuclear energy, appropriately designed to avoid proliferation concerns and to operate in absolute safety, can play a major role in energizing the rest of the world." If you want to understand these issues—whether you agree with this claim or not—you have to hear out this outstanding Senate leader. He has mastered the details and studied the obstacles, dedicating his career to moving them safely out of the way.

Preface

I'm not sure what first prompted me to move from legislator to author on energy issues. Certainly, my thirty-two years of involvement in energy matters was part of it. Statistics about energy needs in the future, the daunting tasks facing underdeveloped nations with few energy resources and massive hunger, the expensive lunacy of American energy and nuclear power policy—all of these intellectual concerns prodded me. In reality, however, it was a more emotional moment that lit the long fuse that has resulted in this book.

One morning in 1990, I woke up to the sound of the television set that my wife, Nancy, had on downstairs. A reporter said something about Iraqi army troops massing on the Kuwaiti border, apparently beginning to invade that nation. I rushed downstairs and watched, for what seemed like a minute but was really an hour or more, while the story unfolded.

Later I heard President George H. W. Bush say that the Iraqi invasion could not stand. America would lead the free world in defense of sovereignty. And so, the drama named Desert Storm began.

Iraq wanted control of Kuwait's oil; in fact, Iraq ultimately wanted a stranglehold over much of the world's oil market. Saddam Hussein calculated crudely, but accurately, that the United States especially would find its economy vulnerable to his whims.

Oil wasn't the reason the United States went to war—Kuwait's sovereignty was. But Saddam Hussein might have had no reason to invade Kuwait if oil had not been a huge part of his calculation. So, the United States sent its troops and its treasure to the deserts of the Middle East, at least in part to preserve national economic security. We had no choice.

I thought about that television scene years later. I remembered the Iraqi troops and the pictures of the men and machines the United States had at its disposal. I thought time after time, "What a tragedy." As we have predicted for decades, stupid public policy has allowed our economy and our troops to be held hostage to our dependence on oil—not just foreign oil, but oil from our land. Tragically, almost no one in political power seems to have a heart to fight for change.

In 1997, as I explain in this book, I decided to use the knowledge I had gathered as chairman of the Senate Appropriations Subcommittee on Energy and Water Development (which funds most nuclear weapons and civilian nuclear energy research in America). I went to Harvard University and gave the "New Nuclear Paradigm" speech in the hope of sparking a new nuclear policy dialogue. As I moved to the chairmanship of the Senate Energy and Natural Resources Committee, I was able to further talk about nuclear issues through a series of hearings on energy, and on nuclear energy in particular. Finally, another dramatic television image on September 11, 2001, confronted me. Troops in Afghanistan. Troops in Iraq again. Soaring gasoline prices. Industry leaving America as natural gas prices ballooned. Protests against wind energy farms, liquefied natural gas (LNG) facilities, drilling in deep waters offshore, opening the Arctic National Wilderness Refuge to drilling, an Alaska Natural Gas Pipeline, and even new dams.

Simply not content with merely holding hearings and proposing legislation, I decided with renewed vigor and a sense of urgency that it was time to broaden my participation in the debate. I decided to write about our energy policy failures, the ironies of our energy policy, and what we must do to put nuclear power back on track in America.

We are willing to spend billions on exotic alternatives, some of which hold virtually no prospect of helping to end our energy dependency. We are willing to dig more coal, risk more miners' lives, pollute more air and streams, and spew more deadly heavy metals into our cities. We are even willing to send our military men and women overseas, where many die and more are wounded.

It takes my breath away that we are unwilling to take advantage of nuclear energy and the billions of dollars' worth of brains that American scientists in the nuclear arena offer us. Why?

I believe two major factors account for our failures. First, many Americans have an irrational fear of anything "nuclear." They are willing to use radiation to fight cancer, detect other diseases, irradiate food. But mention nuclear energy and a kind of strange syllogism forms in the minds of millions of citizens—nuclear means bombs and mushroom clouds; radiation means death. That almost instinctive chain of thought, flawed as it is, gives use of nuclear energy a pretty steep hill to climb with many of my neighbors.

Second, and even more difficult to overcome, has been the policy of deliberate misinformation that opponents of nuclear energy employ with shameless disregard of the truth. Americans already leery of anything nuclear during the height of the Cold War were particularly susceptible to campaigns of misinformation and propaganda. Adding the failure at Three Mile Island and the Chernobyl incident, the misinformation was like kerosene on a fire. In some ways, the propaganda resembled the antifluoride hysteria of the 1950s, in which opponents claimed that putting fluoride in public water systems would lead to widespread death.

While almost no politician was willing to advocate expanding nuclear energy, many politicians made their careers by fighting nuclear energy. Just as bad money drives out good, bad science too often drives out good science.

So, I decided to write this book. My views still get an endorsement of a minority of politicians, if a majority of serious scientists. Some of my fellow senators will think I have lost my mind by speaking too frankly. But it's time for a new seriousness—unless we really like Americans dying in foreign lands, energy prices driving entire sectors of the economy out of business, and dirtier air and water.

Acknowledgments

I would like to acknowledge and thank the following individuals for providing insight, advice, technical expertise, background information, and support: Leigh F. Anderson, Frederic Bailly, Stephen E. Bell, Harold D. Bengelsdorf, Tiffany Berman, Carol Berrigan, Skip Bowman, Michael Cappiello, Gail dePlanque, Nils Diaz, Greta Dicus, Marvin Fertel, Alexander Flint, James Flynn, Chris Gallegos, Paul Gilman, Jean-Claude Guais, Laura Holgate, K. P. Lau, Pete Lyons, Barrie McLeod, Michael McMurphy, Thomas Meade, Samuel Nunn, Hilliard W. Paige, Charles H. Peterson, Edward L. Quinn, Richard Rhodes, Louis Rosen, Michael Schwartz, Eileene Supko, John J. Taylor, and Andrea Welles.

1

My Vision: Reinvigorating "Atoms for Peace"

More than fifty years ago, on December 8, 1953, President Dwight D. Eisenhower offered a prophetic vision to the United Nations that nuclear power could contribute to the betterment of mankind through its peaceful application. At that time there were no commercial nuclear power plants, only prototypes of a submarine thermal reactor and nuclear weapons. The president was proposing ideas on how to "strip its military casing and adapt it to the arts of peace" in order for nuclear technology to provide "abundant electrical energy." In his speech, dubbed "Atoms for Peace," President Eisenhower said:

> The United States knows that if the fearful trend of atomic military build-up can be reversed, this greatest of destructive forces can be developed into a great boon, for the benefit of all mankind. The United States knows that peaceful power from atomic energy is not a dream of the future. The capability, already proved, is here today. Who can doubt that, if the entire body of the world's scientists and engineers had adequate amounts of fissionable material with which to test and develop their ideas, this capability would rapidly be transformed into universal, efficient and economic usage. The more important responsibility of this atomic energy agency would be to devise methods whereby this fissionable material would be allocated to serve the peaceful pursuits of mankind. Experts would be mobilized to apply atomic energy to the needs of agriculture, medicine and other peaceful activities. A special purpose would be to provide abundant electrical energy in the power-starved areas of the world.[1]

We also recognize a second fiftieth anniversary, that of the initial criticality of the first prototype submarine thermal reactor, Mark 1, in Idaho, for a submarine nuclear propulsion plant.[2] Captain Rickover, who was within seven years of hanging his hat at Oak Ridge in 1948, obtained congressional approval to build the naval reactor program and took a nuclear powered submarine to sea.

Recognizing the potential application of the technology developed for the U.S. naval program, President Eisenhower directed Captain Rickover to build and oversee the operation of the first U.S. commercial power plant, the Shippingport Atomic Power Station, near Pittsburgh, Pennsylvania. Having chosen the pressurized water reactor (PWR) as the best technology suited for submarine operations, Rickover proceeded to use it as the basis for the first commercial nuclear power plant. Shippingport's PWR design and original cores became prototypes for the majority of commercial power stations built thereafter. This is the perfect example of the positive synergy between commercial and military nuclear programs.

Marking the more than fifty years that have passed since President Eisenhower's speech and the demonstration of the Mark 1 submarine reactor, I write this book about the progress we have made in our efforts to bring the benefits of nuclear power to mankind. My intent is to outline what went wrong in this noble quest and why, and what we must now do to recover from and repudiate past blunders. It is a book about my vision for a renewed commitment by this nation, and the rest of the world, to the dreams that nuclear energy could help us fulfill. It is also a book about what kind of world our grandchildren could inherit if we fail in making and keeping such a commitment.

For many years now, such a book would have been laughed off the public policy stage. Three Mile Island, Chernobyl, uninformed hysteria about nuclear power's true risks, and irresponsible decisions by policymakers have made nuclear energy into a pariah in some minds. Pseudoscience has been pumped into our culture like bilge water, flushing out the good science. Thus it has been for three decades in the United States.

As we have floundered about searching for a magic solution to our growing energy dependence on foreign and too often hostile sources, we have failed to adequately pursue one of the cleanest, safest, and most reliable fountains of power in energy history—nuclear energy. We have paid an unacceptable price for what can only be called silliness. We have paid that price in billions of wasted dollars due to not basing regulations on sound science, in hundreds of lives lost on the battlefields of the Middle East in our unending quest to maintain a steady supply of oil, and in allowing this nation to be held hostage by our global enemies in any number of policy decisions.

Worse, as we have deliberately limited our potential energy resources by not openly embracing the nuclear option, we now face a future in which global oil production may be already peaking, the price of natural gas will soar, and the promise of options such as solar and geothermal energy remain economically unfeasible in any reasonable timeframe.

We have time to recover from the sins of the past, but we must not delay because the window is closing. We have within our reach, if we will but stretch out our arms, an energy source that can raise standards of living throughout the globe, end our crippling dependence on foreign sources of hydrocarbon fuels, and help clean the world's air and water.

My Vision

In the twenty-first century, nuclear power will be a major contributor to global peace and a better quality of life for both the developed and developing world. My ultimate goal is that in the year 2045, one hundred years after the detonation of the first atomic bomb and the birth of the nuclear age, the world will evaluate the role played by nuclear technologies and conclude that their overall impact was strongly positive.

The world will be a better place in the twenty-first century because of nuclear power. We are already realizing some of the benefits of nuclear technologies today. Nuclear weapons, for all their horror, brought to an end fifty years of worldwide wars in which 60 million people died. On the one hand, nuclear power is providing about 20 percent of the electricity needs of the United States, and on the other hand, many of our citizens enjoy healthier, longer lives through improved medical procedures that depend on nuclear technologies. The standard of living we enjoy today would be lost without reliable, clean, cost-effective electricity. It enables countless technologies— from the computers we use today, to the washing machines that have replaced our old hand-cranked units—making our lives better each day. There are many contributions nuclear technologies could or do offer our society:

- Emissions-free electricity generation and district heating
- Tracing potential cures for major diseases and diagnosing injuries and health concerns
- Destruction of cancer cells through radiation
- Nondestructive testing of the strength of products and components for the manufacturing, construction, and aircraft industries
- Seawater desalination
- Hydrogen production
- Propulsion of submarines, aircraft carriers, icebreakers, and other ships
- Heating to release oil from tar sands and oil shales
- Conversion of nuclear weapons materials to commercial nuclear fuel
- Irradiation of life-threatening bacteria in food
- Spacecraft propulsion by radioisotope thermoelectrical generators and space-based reactors
- Powering nanotechnology's microscopic batteries

- Sterilization of products from bandages to contact lens solution
- Destruction of dangerous radioactive materials in spent nuclear fuel

Nuclear power is safe and sure. Every week, one or two nuclear power plants dock at a major port in America or somewhere else in the world. And these power plants have been doing so for half a century now. Yes, it is true! These nuclear power plants are onboard American submarines. No port (except one in New Zealand) objects to these dockings, no outcries from the citizens occur. Nuclear-powered sub dockings have become completely routine events, and in fact are safer and less stressful for dockworkers than the arrival of ships with cargoes of liquefied natural gas, or ammonium nitrate, or dozens of other much more dangerous substances. No accidents of any kind have ever marred these dockings; no leaks have cleared blocks of cities; no emergencies have been declared. This half century of safety, evidenced by the U.S. nuclear submarine fleet, shows more dramatically than almost any other indicator the safety that can be achieved by nuclear power plants. It is upon the assurance I take from this safety record that I make many of the suggestions for the future of nuclear power in our world and at home in America.

Two words must be part of every discussion on energy alternatives: risks and benefits. No energy source is free of both. Antinuclear groups have focused only on the risks involved with nuclear power. They don't acknowledge its benefits or discuss the solid technical solutions obtained. Unfortunately, their actions do not present a balanced understanding of this complex issue. Energy production, by any technology, represents a trade-off between risks and benefits. The public must have the information to fairly judge both sides of this equation for each type of energy source. With that kind of comparison, nuclear energy fares very well. As the public comes to understand the true risks and benefits of nuclear technology, I believe it will play an increasing role in future domestic and global electrical supplies.

Only a handful of years ago, it was a real challenge to find a headline mentioning the future of nuclear energy. There was little optimism for relicensing, and any talk about a new plant would have been dismissed as childish optimism. We've come a long way since then, to the point where the possibility of new nuclear plant construction in the United States is quite real. But we are not there yet.

I think there's a good consensus today that our nation and the world are facing immense shortfalls in energy, both in the short term and even more so in the long term. I support the renewable options; but, good though they may be eventually, none of them will replace the baseload power that nuclear energy does and will supply in this century. Today there's real enthusiasm for expanded use of nuclear energy.

My vision for a major future role for nuclear energy involves the increasing globalization of the world's economies. I do not believe that the world can move toward peace and harmony unless the large differences between the "have" and "have-not" nations are addressed.

Energy is the lynchpin of economic development crucial to the betterment of mankind. Dr. Denis Beller and Richard Rhodes (Pulitzer Prize winner for his seminal work on the making of the atomic bomb) appreciate the interrelationship between energy and economic progress. Rhodes and Beller write:

> Development depends on energy supply, and the alternative to development is suffering—poverty, disease and premature death—potentiating violence to force redistribution of material wealth. Beyond altruism, considerations of national security require developed nations to foster increasing energy production in their more populous developing counterparts. For safety as well as security, to meet unanticipated natural, ecological and technological challenges, that energy supply should come from diverse sources.[3]

The standard of living for billions of people in the underdeveloped world lags behind that of the developed world by extremely large factors. Reliable sources of electricity underpin the economies of the developed world. Reliable energy sources are key factors determining each nation's standard of living and are certainly one of the prerequisites for modernization in all developing nations. There is a vast gulf in energy usage per capita between Western nations, especially the United States, and the developing world.

Historically, the road to development starts from an almost complete reliance on traditional biomass but leads to the gradual access to electricity and other modern fuels. Electricity will contribute to improving the lives of the world's poor, especially those in the developing countries, in countless ways. The International Energy Agency's study on world poverty demonstrated the link between energy use and poverty. It said:

> Electric light extends the day, providing extra hours for reading and work. Modern cook stoves save women and children from daily exposure to noxious cooking fumes. Refrigeration allows local clinics to keep needed medicines on hand. And modern energy can directly reduce poverty by raising a poor country's productivity and extending the quality and range of its products—thereby putting more wages into the pockets of the deprived.[4]

Global development offers immense benefits to the American people as well. We benefit from a network of global trading partners. These partners help create markets for our high-technology products. But this will happen only if the rest of the world increases its standards of living to levels that closely match our own. And that won't happen unless they have access to clean, reliable, low-cost sources of energy and electrical power. If the United States does not foster global economic development, there will continue to be widespread suffering from poverty, disease, and premature deaths, and an expanding gulf in living standards between developed and developing nations—conditions that will create instability and a potential for regional and global conflicts.

I strongly believe that nuclear energy, appropriately designed to avoid proliferation concerns and to operate in absolute safety, can play a major role in energizing the rest of the world. We can take action now to help shape a world with minimal or, better yet, no nuclear weapons, wherein its citizens enjoy improved standards of living and clean air. At a minimum, nuclear stockpiles can be dramatically reduced. Nuclear materials and weapons expertise can be appropriately protected so that rogue nations will be unable to acquire them. In my view, nuclear power could eventually provide much more than its current 20 percent of U.S. electrical power, as well as clean power for many underdeveloped countries trying to bring more prosperity to their people.

By helping other nations develop nuclear power now, the developing world can avoid construction of fossil fuel power plants that will seriously impact global air pollution. We can help the developing world leapfrog the smokestack era our country went through on its road to becoming the world's most successful democracy and powerful economy. And by assisting those nations with new sources of reliable energy, they can increase their own standards of living. As they, in turn, need and develop high-technology products, we'll have new trading partners who can help each other reach enhanced prosperity.

Facing the Challenges of Nuclear Power

As we begin this new millennium, immense challenges face the world. For example, future generations may confront the threat of nuclear arsenals in many countries. The world will struggle with seriously degraded air quality as the developing countries try to meet their power demands with polluting conventional fossil fuels, such as coal.

In order to confront the challenges of powering the world in the twenty-first century, we must confess that the United States has made some mistakes in the past. I will discuss my Harvard "Nuclear Paradigm" speech (see appendix A) in some detail later in this book, but it is appropriate to summarize in this chapter some of its key points since it presents the main foundations of my nuclear vision for the twenty-first century. This speech explained where America went wrong with its nuclear power program, and what issues must be dealt with to enable us to get back on track and move forward with the expansion of nuclear power.

I used this speech to speak out about how the premises underpinning some of our past nuclear policy decisions were wrong. In 1977, President Jimmy Carter halted all U.S. efforts to reprocess spent nuclear fuel and develop mixed-oxide (MOX) fuel for our civilian reactors in the expectation that such "U.S. morality" would serve as an example to the rest of the world to follow suit. Why was the premise of the decision wrong? Other countries did not follow "our example" because our policy was considered to be both economically and technically unsound. Our failure to address an incorrect premise has harmed our efforts to deal with spent nuclear fuel and the disposition of excess weapons material, as well as our ability to influence international nuclear power issues. By the end of the twentieth century, eighteen countries[5] had developed nuclear fuel cycle capabilities *without U.S. involvement.* We lost our leadership role in the development of safe, proliferation-resistant technologies.

Another premise I refuted in the Harvard speech was the U.S. policy to regulate exposure to low levels of radiation using the so-called linear no-threshold (LNT) model, the premise of which is that there is no "safe" level of exposure. The LNT model forces us to regulate radiation to levels approaching 1 percent of natural background despite the fact that natural background can vary by more than a factor of three within the United States. Why was this another faulty premise? Scientists now believe that low levels of radiation may cause very little, if any, harm. In fact, there are some studies that suggest exactly the opposite is true—that low doses of radiation may even improve health. Whatever the truth is, it is vitally important that we can create rules based on sound scientific knowledge; otherwise, we will continue to waste billions of dollars meeting unnecessary standards.

I cited another bad decision our policymakers made. Based on popularly accepted myth with no basis in reality, the public voiced its fear of purchasing irradiated food products. In the United States alone, each year there are over nine thousand deaths and one hundred thousand serious illnesses from bacteria-contaminated food. Even though the bacteria in beef can be killed by irradiation without any harmful effect on the beef or the consumer, the Federal Drug Administration would not sanction it. I reminded my Harvard audience that were it not for food irradiation, our astronauts would not be able to travel in space, and that scientists had already successfully and safely demonstrated the benefits of food irradiation. But the government's foolhardy position on beef irradiation, which was soon overturned, led to many unnecessary food recalls, illnesses, deaths, and economic losses.

I talked about bad past decisions that haunt us and focused on the decisions we need to make for the future. I argued then and continue to do so today that nuclear power is needed to meet environmental goals. We must revisit our policies regarding fuel-cycle technology. We should move forward with dismantling nuclear weapons and convert the excess inventories of weapons-useable materials first into nonweapon shapes and then destroy them as quickly as possible. My speech was also about redefining our whole waste management strategy. Ultimately, I offered to step up to the plate, lead a new dialogue on nuclear power, and help fund new initiatives in Congress.

The promise of nuclear power cannot be realized unless the military aspects of nuclear systems are controlled. I am a strong advocate of dismantling as many nuclear weapons as possible while ensuring the security of the United States, basing the U.S. stockpile on the threat level to our security. In the process of dismantling these weapons, the inventories of nuclear weapons materials must be converted into non-weapons-useable materials. Technical solutions exist to do so through transforming the plutonium weapons pits into nonweapon shapes and burning the plutonium material in commercial reactors as MOX fuel, as well as blending the weapons-grade uranium into civilian reactor fuel. To this end, I have devoted a lot of energy toward the security and disposition of the huge stockpiles of weapons-grade materials from the countries that were formerly part of the Soviet Union. These are literally matters of life or death, world peace or catastrophic war. Recent evidence suggests that if these materials and/or the brains behind them migrate to rogue nations, we doom the world to a dangerous future.

Before nuclear energy can contribute to our future, the nuclear waste issue must be effectively addressed. Unfortunately, in 1982 we chose to freeze our options and focus only on permanent storage of spent nuclear fuel in an underground repository. We are still on that path today, burying the nation's spent nuclear fuel at Yucca Mountain in Nevada—fuel that still contains 95 percent of its energy content! While a disposal repository is absolutely necessary, it is not so clear that our commercial spent nuclear fuel assemblies should be the material that is disposed there.

Technical solutions for disposing our nuclear waste are at hand, but we must summon the political will to make them happen on a timely basis. There are options for handling spent nuclear fuel that could minimize its risks.

Some type of processing of our spent nuclear fuel, or future reuse of the residual energy in that fuel, may be in our best interests some decades into the future. Some of these options would leave open the possibility that future generations will want to extract the substantial energy remaining in "spent" fuel. In fact, it is quite possible that what we label as nuclear "waste" today may be a critically important resource for future energy generation. The renaissance of nuclear energy increases our concerns about the adequacy of natural uranium for the long term and the need for reprocessing associated with advanced fuel cycles. I am not proposing that we start reprocessing today. However, I am becoming more and more convinced that we should maintain a vigorous and well-funded research program on reprocessing technology alternatives to have them ready "if and when," as well as to keep a seat at the global table when reprocessing issues are discussed.

Transmutation technologies offer the hope of eventually reducing the volume and toxicity of wastes to be placed in a final disposal facility as well as utilizing the residual energy value in the spent nuclear fuel. If we develop a national system of interim storage, as I have been promoting for so long now, we can decide to proceed with the permanent repository while also allowing us the time to explore whether other spent nuclear fuel processing and disposal policies might be more appropriate in the future.

In addition to the need for progress on waste and proliferation issues, the erosion of the educational base and infrastructure in nuclear and related engineering technologies is of serious concern to me. We must breathe new life into university engineering programs and research funding, as well as improve the infrastructure at our national laboratories in order to preserve our future nuclear workforce and America's leadership in the nuclear field. One of the greatest challenges to the nuclear industry in the near future will be maintaining a base of trained workers. In the process of shoring up our workforce and infrastructure, we can revitalize the research and development programs that are essential for the United States to remain one of the world's technology leaders and to develop advanced fuel cycles.

In the near term, nuclear energy must survive in an increasingly deregulated and competitive environment, while maintaining the highest standards for safety that have characterized the industry in recent decades. If we are not careful, deregulation may bring limited near-term benefits to electricity consumers, but may also push nuclear out of the market. The government needs to find the right balance of support mechanisms for sharing the risks and costs of building new nuclear power plants in the deregulated environment.

The nuclear industry will only survive if the regulatory environment is focused on risk-based criteria. Maintaining the pace of reforms at the Nuclear Regulatory Commission (NRC) is crucial. We must also address the problems associated with the use of the LNT model for assessing radiation risks by exploring the scientific basis for radiation standards, which may help set better regulatory standards and address public fears about radiation. I believe that progress on waste issues and a better understanding of the risks of radiation, which have been overstated by the current regulations, will go a long way to improving the public's acceptance of nuclear energy.

Security and safety issues are intertwined in the public's mind and must be dealt with to the public's satisfaction in order to engender the level of public confidence necessary for nuclear power's renaissance. The terrorist attacks against the United States on September 11, 2001, changed many people's conceptions of the threats we face, and some question the safety of nuclear plants in light of those attacks. I concur that it is appropriate that we carefully evaluate the safety of all major nodes of our critical infrastructures. However, we must remember that nuclear plants are probably the most hardened commercial structures in the world. The antinuclear factions' arguments that we should abandon any facility that cannot be totally off-limits to terrorists are fallacious. If we heeded their logic, we would need to abandon trains, chemical plants, airplanes, and skyscrapers.

Blemishes on nuclear power's safety record must be addressed and placed into proper context. The Chernobyl accident in 1986 happened at a badly engineered reactor, because of a human operating error, under forbidden and secret circumstances in a poorly regulated industry, and under a communist regime. Tragically, thirty-one people died and many more suffered unnecessarily from thyroid disease because of the lack of an appropriate emergency health care response. This type of accident would not happen to Western-designed reactors, in a Western regulatory environment. The Three Mile Island accident was also caused by a human operating error, but it was not the "China Syndrome" that the antinuclear movement and Hollywood have painted such an event to be. No one was seriously injured—safety measures worked and radioactive material releases to the atmosphere were minimal. Unfortunately, the costs to the industry were large indeed and set nuclear power development back many years.

The human safety record of nuclear power in the United States is exemplary, attested to by the outstanding record of the industry as well as our nuclear navy. Incorrect public perceptions about the safety and security issues must be dealt with in a manner so that the public can adequately assess the relative risks and benefits of this clean energy source vis-à-vis the major alternatives—coal, oil, and gas, as well as renewables.

Conclusion

We have the technical expertise to realize a better world. All around are examples of well-engineered systems that surmount daunting barriers to provide us with significant benefits—from jet planes to bridges and skyscrapers. For any of these systems, there are significant risks. But when those risks are evaluated and countered with solid science and engineering, we have great confidence in their performance.

While public debate on nuclear technologies was stalled in the latter part of the twentieth century, there are now indications of a reawakening. There is a new awareness that the premises of many past policies and positions were wrong and that the environment and economy in the United States and the world will increasingly be dependent upon nuclear energy—and the better for it. In order for this to come to pass, energy policies need significant revisiting and revitalizing.

The United States is in a unique position with its resources and technical expertise, if it can summon the will to pursue sustained political and policy leadership to promote a more peaceful and just future. Our nation is poised to meet not only our needs for electricity with clean, low-cost generation, but we can reach out to other nations and help them.

America has gone a long way in keeping its promise of expanding the peaceful use of nuclear power around the world. Indeed, our country is taking steps to make good on President Eisenhower's promise to rid the world of nuclear weapons. It is time to reinvigorate Eisenhower's dream by addressing the roadblocks to the further development of nuclear power while moving toward Eisenhower's dream of a nuclear weapons-free world. We must think of our challenge not as putting the nuclear genie back in the same bottle, but as building a new, larger bottle in which to contain the genie. We can do that, and the opportunity to do so exists right now.

This book aims to outline what went wrong, why, and what to do to recover from and repudiate past blunders. I devote these next chapters to how we can tackle the issues of proliferation, waste management, safety and security, public acceptance, regulatory culture and standards, workforce and infrastructure maintenance, uranium resource adequacy, and the future direction of nuclear technology development. But first, I want to talk about my family and how I became involved in nuclear energy issues over the past thirty years.

2

The Road to Leadership

Growing Up in My America

I was thirteen when the United States dropped the first atomic bombs to end World War II, but beyond that, anything nuclear was foreign to me. Growing up in Albuquerque in that era, my awareness of energy sources was limited to the electricity it took to run my father's grocery wholesale business and the gasoline it took to keep the trucks running. A typical young American boy, my life focused on sports, school, and working for the family business, and certainly not on any notion that revolutionary technologies were being developed on a mountain less than one hundred miles north of my family's home.

However, my solid belief in nuclear power as a necessary and desirable energy source for the future was from the beginning rooted in my upbringing as a first-generation American and the beliefs that my Italian parents instilled in my family and me. So this book is also the story of how my family came to America and the path I walked to become one of the strongest proponents of nuclear energy in the Congress of the United States.

I am a first-generation American, born in Albuquerque, New Mexico, on May 7, 1932. The remarkable story of how I came to be born in America goes back to the early days of the twentieth century. In 1906, when my father was fourteen, both he and his twenty-one-year-old brother, Anthony, were brought by their Uncle Lorenzo Gradi from Italy to the United States. My father was a farm boy from the little town of Lucca in northern Italy, in the province of Tuscany, where the people proudly refer to their dialect as *el verro italiano*, the true Italian. My father was the last born of this Italian family, and he had a beautiful name that in Italian means "Little Angel." He was the youngest child in the family and his mother had died in childbirth.

Uncle Lorenzo decided on one of his frequent trips back to the homeland that my father should have a chance to go to America. He told his thirteen-year-old nephew that in the next year, he would sail back to Italy and bring him to Albuquerque and the great land of opportunity. When Uncle Lorenzo returned as promised, Cherubino was frightened to death about the idea of leaving home for America. His older brother, Anthony, yelled out, "If you won't go, I will!" After some negotiations, both Cherubino and Anthony headed off to their new life a month later and the opportunity to work in the business created by Uncle Lorenzo.

Before the turn of the twentieth century, Uncle Lorenzo, who spoke little or no English, owned a mercantile store in Albuquerque. The store was named, strangely for an Italian, "Montezuma," after the ancient leader of the Aztecs. No one knows why it was called the Montezuma Mercantile Company. We still have a photograph at home of the four-story building and its varied wares such as cowboy boots, Italian imported oil, and everything in between.

When Cherubino and his brother Anthony arrived in Albuquerque in 1906, the city of Albuquerque only had a population of about 4,500 people, compared to its current population of more than 650,000 in the metropolitan area. The principal language was Spanish, although there were plenty of English-speaking people living there. My family learned Spanish, English, and Italian, but my father, who never went to school because the laws at the time did not require anyone over the age of fourteen to attend, never learned English well.

Only five years later, in 1911, the opportunity of a lifetime arrived. When my father was nineteen, Uncle Lorenzo decided to retire and go back to live out his life in Italy. He sold the business to his nephews, and as time passed, the business was converted to a wholesale grocery company that furnished employment for me and one of my sisters.

This is a classic story of Italian immigrant life and the American dream. I, along with my three older sisters and sister Roseanne, who was twelve years younger, attended Catholic schools, and lived right near our school in downtown Albuquerque. As I grew up working in my father's business, I learned two important things that have served me well in the intervening sixty years—how to speak Spanish, and how small business capitalism really works. Capitalism was a daily event in our household. My father worked every day, including Saturday until noon, and even Sunday until about 10 A.M., stopping just in time to go to church. As a family we ate all our meals together and our father returned every day at both lunchtime and in the evening so that we could gather around and eat. As a consequence, as you might guess, we shared a lot in the ups and downs of the business. I knew when the business was succeeding because my father and uncle would bring a new truck home and share the excitement with us, particularly if it were a big truck that had more wheels than the previous truck!

Family affairs dominated our lives. We were good students and our folks were tough disciplinarians. School was the most important thing in our lives outside of family and church, and we were expected to do well. I never dreamed of being a politician, but I found myself interested in people and problems; and there's no question that whatever group of people I belonged to, I sought to be its leader. I was president of my class in the ninth, tenth, and eleventh grades. I graduated from St. Mary's High School in Albuquerque in 1950 and headed for the University of Albuquerque, a school now closed, where I was class president as well.

In 1953, I transferred to the University of New Mexico and earned an education degree the following year. Following a successful college pitching career, I became a pitcher for the Albuquerque Dukes, a farm team for the Brooklyn Dodgers. A year of trying to get my curve ball over the plate, with little luck, persuaded me to use my education degree to become a mathematics teacher at Garfield Junior High in Albuquerque. Then I concentrated on obtaining a law degree from the University of Denver, which I did, in 1958, and married my sweetheart, Nancy Burk, that same year. We became the proud parents, over time, of two sons and six daughters.

Although an adult in my mid-twenties, during this period I still gave no thought to either politics or nuclear power. While I certainly knew that the atom bomb over Japan ended World War II, I did not then understand that the atom would later be harnessed to provide power to propel ships and to generate electricity. I was not even aware that the United States' first civilian electricity generating nuclear power plant began operating at Shippingport, Pennsylvania, at the end of 1957, shortly before I earned my law degree. I must confess that at that time I had very little understanding of how a nuclear power generating plant worked.

I knew that the atomic bomb had been developed at the "Secret City" of Los Alamos, on the Pajarito Plateau, just one hundred miles north of my home. We vacationed in the Jemez Mountains, near the village of Jemez Springs, just forty miles from Los Alamos, where we would stay in a cabin for two weeks each year, along a beautiful small stream. Our dad would occasionally take us on a Sunday drive in our old Packard automobile up the Jemez Mountains to the gates of the mysterious closed city almost two hours away. And all we knew was that guards were there, and we were to make a U-turn to head back to where we came from, and "move on, please!" Father used to tell us, "Big things are happening up there in Los Alamos. America's doing big things there."

It was many years before I learned how right my father was. During the early 1950s the United States and the Soviet Union began in earnest the Cold War. Much of the United States' nuclear weapons work then was actually taking place at Los Alamos, the site where the first weapons were designed and built during the 1943–1945 period. The first man-made nuclear explosion, the Trinity bomb test, took place in July 1945 in the Jornada del Muerto (Journey of Death) region about 125 miles south of Albuquerque. Maybe it was, upon reflection, this knowledge that made me determined to see nuclear power harnessed for human development.

By the time Nancy and I married, the atom was well on its way toward providing electricity for millions of Americans and propelling nuclear submarines. While I did not know it at the time, expanding the valuable work done at Los Alamos, helping to solve the concerns of the nuclear industry, and ensuring the success of our nuclear navy would become a thirty-year project when I became a U.S. senator.

My Path to Politics

How did someone who had never shown an interest in politics get involved?

Strange as it may seem, I entered into politics as the result of a "dare," in 1966, and my stubbornness. I complained about the state of city government to five of my best friends who finally said to me, "Either run for office, or quit bitching!" So I decided to run for the Albuquerque City Commission. Even though my friends then tried to change my mind—warning me that if you get in, instead of having friends, you'll have enemies; instead of moving ahead, you'll go backward—I ran for office. I was elected a commissioner of the city of Albuquerque that year and served four years, three of them as chairman ("mayor") of the commission. In 1970, I became the Republican nominee for governor of New Mexico, but lost to Bruce King, who went on to serve three terms during the next two decades. I returned to practicing law; but not for long.

Six years and four months from the time of the famous dare, I was elected in 1972, at age forty, to the U.S. Senate—the first Republican senator in thirty-eight years from the state of New Mexico. While we thought at the time that we might be too poor to run for the Senate, a sizeable court victory, and Nancy's ability to pinch every penny and manage our way through the campaign, allowed us to struggle through what was, even then, a very expensive run for the Senate.

The campaign experience reinforced in me something my whole upbringing led me to believe—that for almost all human problems, human solutions can be found. I came to this point of view, as you might expect, because of my Catholicism and in part because I had seen my father and our family meet challenge after challenge by simply managing our affairs logically and with discipline. I came to the Senate with a belief that all problems could be solved, and that if we were not solving them, it did not mean that we should accept failure. I believed, rather, that it was our fault as humans that we were not solving them.

A wonderful story illustrates this belief. When I was the mayor of Albuquerque, and father of eight children, I was conducting a town hall meeting with a group that favored population control, which was all the rage at the time. I was never an advocate of population control, because I thought that the problem was not one of too many people but rather one of management. In response to their criticism of me for having more than the recommended two children per family, I pulled out a bunch of papers in my pocket and told my critics that I was right in line with their rules. I said, "Here's a letter from one of my sisters who's been a nun for twenty years. She said I could have two children for her. Here's one from my other sister; she's already past the age of childbirth and she said I could have two for her. Here's one from my third older sister who has never married and says she is never going to have any children so I could have her two. So, those six plus the two I am allowed to have equal eight, making me very consistent with your goals." It worked very nicely, and they did not bother me about that again!

When I joined the Senate in 1972, I quickly honed my legislative skills. In a quiet but effective revolt in 1974—my second year in the Senate—I joined with other future legislative leaders like Sam Nunn of Georgia and J. Bennett Johnston of Louisiana, both Democrats, to lobby for increased funding and greater access to the committee deliberations for freshmen senators. In 1977, I once again joined prominent young Democrats—in a bipartisan effort that I have tried to make my hallmark approach—to tackle energy problems in the nation. My initiatives to create a synthetic fuel corporation and to boost alternative sources of energy both were signed into law by President Carter.

The passage of the Inland Waterways Bill, S.790, in 1978, and signed into law by President Carter, pitted me against one of the most powerful senators of my time—Russell Long of Louisiana, both a policy and parliamentary master. My bill, called an impossible dream by most observers, was designed to impose a new fee on barge operators and owners on those waterways built by the United States. Senator Long, of course, opposed this new fee, arguing that it would hurt his constituents who benefited from the subsidized construction and rehabilitation of locks and dams along such waterways as the Mississippi River. The notion of a new fee was an old one, and the bill was considered dead upon introduction. The difference was my legislative approach. In the end, the bill passed, became law, and thus began the stories of my legislative abilities. The whole story is now a book by T. R. Reid, *Congressional Odyssey*, still used in colleges around the country as the classic account of practical legislating.

The "logical but human" approach has served me well in the Senate, I believe. I joined the U.S. Senate Budget Committee during my first year in the Senate because it offered a chance to tackle the seemingly intractable issue of federal budget deficits. The committee was brand new, part of a massive congressional overhaul of federal budgeting, and had wide-ranging jurisdiction over the federal budget in all its aspects. I moved up in seniority fairly quickly and was propelled into the chairman's seat in 1981, after the Republicans captured control of the Senate as part of the "Reagan Revolution." I was regarded as a deficit hawk—willing to support real restraint and reform of entitlement programs, like Social Security and Medicare, and even tax increases—in order to stem the red ink of the 1980s. I worked to perfect the 1985 Gramm–Rudman–Hollings deficit reduction effort, which, while failing in some sense, succeeded in making deficit reduction the prime fiscal focus of the Congress. This led to the 1990 budget deal, which raised taxes in order to cut the deficit, much to the chagrin of my more ideological Republican colleagues. My spending cuts infuriated the political Left and my willingness to raise some taxes discomforted some on the Right. I am proud of the more than twenty federal budgets I helped to usher through the Congress as the senior Republican on the committee, especially when, in 1997, we passed the first balanced federal budget in four decades.

While federal budget matters dominated my work in the Senate, I was lucky in obtaining another assignment in 1977—the Senate Interior and Insular Affairs Committee, a committee with vast responsibilities for land, water, and other resources in New Mexico and the western states. Soon, the Senate expanded the jurisdiction of the committee and renamed it the Senate Energy and Natural Resources Committee, adding most of the energy policy issues to its portfolio. Thus, energy in all forms became a central issue for me as a U.S. senator.

Even before I became chairman of the Senate Energy and Natural Resources Committee in 2003, I addressed questions of nuclear power, nuclear weapons, and nuclear waste. After all, I was a senator from a state that had the longest history of work with nuclear materials of any state in the union. One cannot be a senator from my state and remain ignorant of nuclear issues. Los Alamos National Laboratory, Sandia National Laboratories, and the Special Weapons Center at Kirtland Air Force Base in Albuquerque demanded huge attention from many serious policy perspectives.

My Introduction to Nuclear Issues

My formal introduction to nuclear power began immediately after I became a senator, as it became obvious that I would need a crash course in order to represent the national laboratories in New Mexico. Fortunately, it was at about this time that I met Dr. Louis Rosen, director of the High Energy Physics facility at Los Alamos, and now a friend of more than thirty years. Dr. Rosen began working at the laboratory in 1944 during the Manhattan Project. In 1972, he saw the results of more than four years of great effort come to fruition, with the powering up of the Los Alamos Meson Physics Facility (LAMPF), a premiere nuclear physics facility for the nation and the world.

Dr. Rosen briefed me several times in great depth and made available to me many other members of perhaps the world's most respected nuclear facility in the world. He asked to be appointed to a United States–Soviet Union committee whose task it was to review independent proposals on nuclear matters involving the two nations. In this role, he was also able to keep me up to date on economic and technical developments in the former Soviet Union, including the growing stockpiles of weapons of mass destruction there. At the same time, I was receiving frequent briefings from Sandia Laboratory specialists and other key nuclear scientists throughout the United States. My modus operandi therefore was to set up advisory groups to help inform me about the issue at hand. I was able to access government experts, university researchers, and physicists with the widest imaginable backgrounds. My education moved along quickly indeed and I believe the process served me well.

A formative event in the development of my nuclear policies came when Congress began to debate the authorization of the Waste Isolation Pilot Program (WIPP) in the state of New Mexico. It was then that I came to understand how important it is for a potential host state to have an active consultation and cooperative role in, but not veto power over, siting decisions. When it came time to pass the fiscal year 1979 Defense Authorization, I had to do yeoman's work in convincing the conferees that New Mexico must play a key role in the decision-making process. I am grateful to Senator Carl Levin for his assistance in what proved to be a winning position.

In the early 1980s another opportunity presented itself to me in the Senate. I became a member of the Senate Appropriations Committee and its Subcommittee on Energy and Water Development, which funds, among other things, most of the United States' nuclear weapons programs. Now, with my positions on both the Energy and Appropriations Committees, I was able to focus more than ever on matters nuclear. The briefings and hearings increased in number and intensity as I helped to develop spending plans for the nation's nuclear complex. And so, by the mid-1980s, I felt comfortable dealing with the interrelationships between nuclear power, nuclear weapons research, weapons policy and nonproliferation issues, and the intersection of these issues with those of our country's energy demands and environmental concerns.

I thought increasingly about the ramifications of these issues each time we confronted the questions of sustainable, economic energy supply, the plight of the underdeveloped world, global population increases, and America's increasingly dangerous dependence on imported petroleum. Was it true, as some claimed, that the Western nations would be required to cut back on their economic growth in order to leave room for other countries to grow? Had global economics all of a sudden become a "zero sum game," where progress in one region could only come at the expense of other regions? I was aghast at such conclusions. Nothing would thrill me more than to have all the underdeveloped nations become more prosperous and healthier, with jobs and real futures. However, it seemed irrational and unnatural that only restraint of growth in the Western nations could lead to global growth. The more I considered these issues in the ensuing years, the more it became obvious to me that we had, right in our hands, developed in America, by Americans, the solution to powering this growth in a sustainable way—nuclear power!

The more I had facts presented to me from good people—about the success of nuclear power, how safe it really was compared to all the other sources of power, how clean it was in terms of the ambient air—it became more and more obvious to me that it was a responsibility of mine to find out what had gone wrong, why we had stopped moving ahead. I knew that if more science, energy, and resources were put into it, we could make nuclear power plants even better and safer, although it already seemed to me that they were the safest.

I visited countries like France and found out that they were using nuclear power in abundance, and that here America sat—squeamish, frightened, big America—worried about the technical problem of what to do with the spent nuclear fuel coming out of the fuel cycle at its tail end. As Ivan Selin, chairman of the Nuclear Regulatory Commission, said, "You have to admire a country like France that has two kinds of reactors and hundreds of kinds of cheese, as opposed to a country like us that has hundreds of kinds of reactors and two kinds of cheese."

Returning to the evolution of my involvement in nuclear issues in the 1980s, I was alerted to the fact that the U.S. uranium mining industry was about to face hard times ahead. I began to think about issues related to the uncertain job outlook for New Mexico's uranium miners. At the beginning of the 1980s, the market price of uranium entered into a decline partly as a result of the diminished demand outlook for nuclear power following the Three Mile Island incident and partly as a result of the increased availability of competitive, lower-cost foreign supply. At the time this did not at first seem to me to be a nuclear power issue so much as a labor issue. Until then the United States had been the Western world's leading producer of uranium, and New Mexico was the nation's leading producer.

My interest in nuclear issues broadened considerably as a result of my exposure to the nuclear programs in other countries. In 1985, as chairman of the Senate Budget Committee, I was part of a congressional delegation, comprised of Senators Bob Dole, James McClure, Daniel Moynihan, Daniel Evans, Bill Cohen, and Pete Wilson, traveling to the Far East. We went to Japan, Taiwan, Hong Kong, South Korea, and the People's Republic of China. While the purpose of this trip was to discuss trade issues, I took the opportunity to discuss energy issues as well. I learned that nuclear power plants provided 14 percent of Korea's electricity and the active involvement U.S. industry had helped South Korea develop its nuclear capacity. In Taiwan I attended briefings about the extent of nuclear power's contribution to its electricity grid—4,000 MWe of capacity representing over 30 percent of its total capacity, with eight more nuclear units planned to be online by 2000. I appreciated that Taiwan was a significant nuclear trading partner of the United States with its commitment to purchase 70 percent of its enrichment services from the United States and to involve U.S. industry in Taiwan's nuclear construction program. It was clear to me that these countries needed a broad energy policy and that several of these countries could be potentially large purchasers of U.S. nuclear fuel.

Over the course of my Senate career, I rarely traveled overseas, but when I did, I tried to make a small contribution to the country I was visiting whenever I could. I remember well, on my visit to South Korea, how important a small gesture on the part of an American official could be to the local community. Our delegation's leader, Senator Dole, was scheduled to preside over the dedication of the new Nancy Reagan wing of the Han Mi Hospital. However, due to concerns over an exhausting travel schedule and protocol issues expressed by the U.S. Embassy in Seoul, Senator Dole was advised not to go to the hospital dedication as planned. My aide, K. P. Lau, traveling with my wife Nancy and me, was alerted to the Koreans' dismay over the turn of events and suggested that Nancy and I go in place of Senator Dole. With a last-minute decision in hand, we needed a police escort to be able to get through the traffic on the thirty-minute trip in order to arrive at the appointed hour. To our astonishment, the entire town of Sun Nam was lining the streets to welcome and thank the Americans for their generosity in funding the hospital wing. Thanks to K. P. Lau's advice, we made the right decision.

During the next several years I found myself increasingly involved in domestic uranium mining and associated nuclear fuel issues, which drew me more and more into the broader issues related to the nuclear power option. Dealing with these issues led my staff—including key people such as Paul Gilman and K. P. Lau—and me to become more deeply involved in the question of restructuring the uranium enrichment enterprise.[1] By the end of the

1980s, I found myself continually involved in trying to weave legislation to enable the suppliers and the utilities to reconcile their nuclear fuel supply differences. Our efforts in this regard finally culminated in October 1992 in the passage of the Energy Policy Act (EPACT).

With the dissolution of the Soviet Union, I seized the opportunity to deal with a broader range of nuclear issues, moving into the nonproliferation arena. Along with Senate "greats" Sam Nunn and Dick Lugar, we created the "Nunn–Lugar" programs to deal with the materials, scientists, and closed cities from the former Soviet Union's nuclear weapons programs. We passed the legislation to enable the creation of the Highly Enriched Uranium Agreement to blend down the material from dismantled nuclear weapons from the former Soviet Union countries' arsenals to be burned in civilian reactors. Tackling the plutonium conundrum, we crafted a proposal that ultimately led to the acceptance of a bilateral agreement between Russia and the United States to dispose of both of their excess weapons-grade plutonium stockpiles.

Of great interest to me over the years were the scientists who would come to talk to me about projects of significance, some of them only peripherally related to nuclear matters. Take for example the great American Genome Project that started mapping the human genome sequence. From the standpoint of Congress, the Genome Project started in my office—that is, the federal government's decision to fund it. It started in 1986 with a scientist from the Department of Energy (DOE), Dr. Charles DiLisi, who had left the National Institutes of Health (NIH) because of frustration over their reluctance to do the mapping. Dr. DiLisi convinced me that if the NIH did not want to find funding, then I should introduce legislation for the DOE to do so. One of my motivations for becoming involved was that it needed the technology and computer facilities that were available at Los Alamos. The late Senator Lawton Chiles (later Florida's forty-first governor) and I funded it for two years before the executive branch finally got on board. This project demonstrated to me the political power and leverage of the congressional appropriations process. However, it was also clear to me that it could be used either positively or negatively.

Offering a New Nuclear Paradigm

It seemed that in the waning decades of the twentieth century, America had a system that put the future of nuclear power into the hands of the courts and, as a result, tied nuclear power up in knots. I am so grateful that I was inspired to thoroughly investigate the issue, and finally, to deliver my "Nuclear Paradigm" speech at Harvard in October 1997. It is still, I understand, a storied speech on nuclear power and how federal government policymakers created the current morass.

Steve Bell, my chief of staff, Dr. Pete Lyons, my science advisor, and Alex Flint, who was then my staff director for the Appropriations Subcommittee on Energy and Water Development, deserve much of the credit for putting "the bee in my bonnet" to give the Harvard speech. I have been very fortunate since then to have been assisted not only by Dr. Lyons, Mr. Bell, and Mr. Flint, but others on my staff who are superb in this field. We hashed out the ideas, argued about portions of it, until we thought we had it just right. I recall rather vividly that one of the views that I was most worried about promoting was how very foolish we were for not irradiating our food. I asked my staffers, "How much do you want me to bear in one speech? I will be pilloried for wanting to irradiate our food." They replied that I need not worry since we were already doing it! They reminded me that the reason our astronauts could eat food—safe food—for so long in space was because the space rations were irradiated. I took their word for it and left the food irradiation argument in my speech. Interestingly, it was not more than a week after the Harvard speech that the Federal Drug Administration approved the consumption of irradiated beef by American consumers.

At Harvard I introduced concepts for a new nuclear paradigm. In this paradigm, our nation's strategies in both civilian and military aspects of nuclear technologies would be coordinated. Proliferation risks would be reduced and nuclear energy would be retained as an option to meet our future energy needs. While there has been immense progress toward this new paradigm since that speech, especially in the areas of nonproliferation programs and serious research on better approaches to waste management, it is time to reinvigorate my promise to provide leadership and to advance the agenda.

The New Nuclear Dialogue Takes Shape

In the late 1990s, President Clinton outlined a program to stabilize the U.S. production of carbon dioxide and other greenhouse gases at 1990 levels by some time between 2008 and 2012. However, it was clear to me that these goals would not be achievable under Clinton's energy policies without seriously impacting our economy. What President Clinton should have proposed was an increase in reliance on nuclear energy to meet our environmental goals.

President George W. Bush took a more pragmatic approach to stabilizing greenhouse gas emissions. In February 2002, the president announced several nonregulatory initiatives intended to reduce those emissions. Among other things, these initiatives included reducing the ratio of emissions to economic output by 18 percent over ten years, thereby reducing the carbon intensity of the U.S. economy from an estimated 183 metric tons of carbon per million dollars of gross domestic product in 2002 to 151 tons by 2012. President Bush emphasized the important role of nuclear energy in achieving these goals.

Ironically, the required technology to meet our environmental and energy goals has been with us for several decades, ready and waiting to meet the challenge. We have developed the next generation of nuclear power plants— which have been certified by the Nuclear Regulatory Commission and are now being sold by American companies overseas. They are even safer than our current models, if that is possible. Better yet, we have technologies under development, like passively safe reactors, high temperature gas reactors, and advanced liquid metal reactors that generate less waste and can be more proliferation resistant.

When the Republicans regained control of the Senate at the beginning of 2003, I became chairman of the Energy and Natural Resources Committee. I gave up the budget chairmanship because I had done everything that could be done from the standpoint of budget leadership, and, equally important, the country was facing the need for a new energy policy. At the time there was a list of highly controversial outstanding issues related to regulating the electricity industry, issues that had splintered the Republicans and stymied the Congress since the early 1990s. I remember that we gathered together in one room on April 29, 2003, for a horse-trading session on the energy bill. When we entered that room we were as diverse as a group of chickens that had just gotten scared to death by a hawk. We were all wondering where we were going. Two hours later, we left with unanimous agreement, or so we thought.

The Senate formally took up the Energy Policy Act, S.14, in May of 2003, and planned to set aside one week to complete the bill. The bill had hundreds of amendments pending, ranging from automobile fuel efficiency standards to financial support for new nuclear power plants. In the previous Congress (the 107th Congress), the Democratic-controlled Senate and the Republican-controlled House each had passed bills that differed significantly and that were never reconciled in conference committee. While we finally got S.14 to the floor on July 28, debate on the various amendments was such that during the afternoon of July 31, it became clear that the bill was too contentious for resolution in the limited time available. We decided that the only strategy that would result in an acceptable bill being passed during that particular Senate calendar year was to approve the Senate-passed version of the energy bill (H.R.4), which we did, eighty-four to fourteen, and send it to Senate–House conference in the fall. The bill that we passed served as a vehicle to get to conference. My epilogue describes the twists and turns the legislation has taken. Suffice it to say, however, I will not rest until Congress passes this vital legislation, putting America on the path to a cleaner, healthier, sustainable, and self-sufficient energy future. This is my commitment to my fellow Americans.

3

The Energy Highway

My vision for energy policy in the twenty-first century rests upon the recognition that by the end of the century there will be three primary supply sources on the energy highway: nuclear power, clean coal, and renewables. While hydrogen will largely replace oil in the transportation sector, it will most likely be produced in large quantities by "water cracking" in high-temperature nuclear power reactors or by coal gasification plants.

The inescapable fact is that energy consumption may double in the next fifty years, as the world's population is projected to increase from 6 billion today to between 8 and 12 billion by mid-century.[1] A recent conservative estimate by the U.S. Census Bureau predicts world population will reach 9 billion by 2050.[2]

The grim picture is that today's sources of fossil fuels are universally agreed to be pollutants and contributors to greenhouse gas emissions and acid rain, particularly coal and oil. Their percentage of energy supply is already approaching their peak, and possibly has already reached that point in the case of oil. Many believe that world reserves of oil will be seriously depleted in forty-five years and natural gas reserves in sixty-five years. Production of oil and natural gas is projected to peak around 2020.[3]

Especially worrisome are the geopolitical, resource adequacy, and environmental issues surrounding the continued use of fossil fuels. Europe and Japan are already in the resource grip of Russia and the Middle East. We in the United States also know that our country is overly dependent upon foreign countries for increasing supplies of oil, and further, that the same may soon be the case for gas supplies.

These resource, security, and environmental concerns lead me to conclude that nuclear power could be a heaven-sent alternative to polluting fossil fuels for energy supply in this century. However, it is also a given that renewables and clean coal must also play their part in fueling the insatiable need for more energy that this world will continue to face.

To understand how nuclear energy fits into this picture, we must first look at the overall requirements for a diverse supply of energy and how much electricity the world will need. This chapter provides the context for energy supply and demand both in the world and the United States, while chapter 4 is devoted entirely to nuclear power.

World Energy Requirements

Vocabulary

Before discussing energy supply and demand, I would first like to introduce and explain some of the subject's basic terminology, in words that are not generally in everyday public use. Let me start with energy, which may be defined as the capacity to do work, and for which the basic unit of measurement is the joule,[4] after the noted eighteenth-century English thermodynamicist James Prescott Joule. Because the joule is a very small unit, it is common practice to use the following "shorthand" amplifying prefixes: mega (million), giga (billion), tera (trillion), peta (million billion), and exa (billion billion). For example, one barrel of oil contains about 6 gigajoules, or 6 billion joules.

The next important energy unit is power, which is defined as energy per unit of time. The term for the basic unit of power is the watt (W), named after the Scottish inventor of the steam engine, James Watt, and which is equal to one joule per second. Common units in everyday use for power are:

kilowatt (kW)	=	one thousand watts
megawatt (MW)	=	one million watts
gigawatt (GW)	=	one billion watts
terawatt (TW)	=	one trillion watts

A common unit of energy in electricity is the number of watts used in one hour. This use of a TW of power for one hour is one TWh. Electricity demand can be expressed either in the average power used in watts or with energy consumed in some period of time in watt-hours, or Wh.

Table 3.1 presents common energy conversion factors used by engineers and economists on a regular basis. They are helpful in interpreting various energy projections that are not always given in similar units. One particularly convenient and popular energy unit shown in the table is the quad, which is the shorthand for a million billion or quadrillion British thermal units (Btu).

Table 3.1. Energy Conversion Factors

From One:	To:	EJ	Gtce	Gtoe	Tcm NG	Quad
Exajoule (EJ)		1.000	0.033	0.022	0.025	0.948
Billion MT coal equivalent (Gtce [a])		30.300	1.000	0.675	0.761	28.720
Billion MT oil equivalent (Gtoe [a])		44.900	1.482	1.000	1.128	42.559
Trillion cubic meters natural gas (Tcm NG)		39.800	1.314	0.886	1.000	37.725
Quadrillion British thermal units (Quad)		1.055	0.035	0.023	0.027	1.000

Source: National Academy of Sciences, Cooperation in the Energy Futures of China and the United States (Washington, D.C.: National Academies Press, 2000), 92.

One quad is the equivalent of 0.293 TWh (thermal) or approximately 0.1 TWh (electric)—the difference is due to the conversion losses in the generation of electric power. One Btu is the amount of energy required to increase the temperature of one pound of water (one pint) by one degree Fahrenheit, about the amount of heat produced from burning one common match.

Energy Demand and Supply Outlook

Energy Demand Overview

To put energy policy into context, it is important to gauge just how much energy is consumed today, and how much will be needed in the future. However, comparing energy and electricity forecasts tends to be confusing because the start dates in the many studies range from 2000 to 2003 and the end dates range from 2020 to 2050. The U.S. Energy Information Administration (EIA) has extended the end date of its latest forecasts to 2025. The International Energy Agency (IEA) goes out to 2030. The World Energy Council (WEC)[5] takes its projections out to the middle of the century, to 2050.

The most up-to-date analyses available for this book show that in 2002, U.S. total energy consumption was approximately 98 quads.[6] World energy consumption stood at 404 quads in 2001.[7] Table 3.2 presents the EIA's forecast for world energy demand, breaking it down by energy source.

The EIA projects that world energy consumption will increase from 404 quads in 2001 to 623 quads by 2025. Most of this increase is projected to come from fossil fuels and very little from nuclear power. It forecasts an annual demand growth rate of 1.8 percent. EIA estimates that the fastest growth, 5.1 percent, is projected for the nations of developing Asia, including China and

Table 3.2. World Energy Consumption (Quads)

Energy Source	Year		Growth Rate (%)
	2001	2025	
Oil	156.5	245.3	1.9
Gas	93.1	156.5	2.2
Coal	95.9	140.2	1.6
Nuclear	26.4	30.4	0.6
Renewables	32.2	50.4	1.9
Total	403.9	622.9	1.8

Source: DOE/EIA, *International Energy Outlook 2004*, DOE/EIA-0484 (2004), April 2004, table A2.

India. EIA projects considerably lower growth in the developing nations, averaging only 1.2 percent.[8] While the annual percentage growth rate forecasted by both the IEA and the EIA are comparable, the EIA's world energy forecast has dropped since the previous year and is now quite a bit lower than the IEA's, which predicts energy consumption will grow to 650 quads by 2030.

The IEA forecasts world energy demand growth from 392 quads in 2000 to about 650 quads by 2030, an overall increase of about 1.8 percent per year, a rate that is similar to the EIA growth rate. Approximately 30 percent of the growth is forecast to take place in the developed countries, that is, in the Organization for Economic Cooperation and Development (OECD) member countries. With a slightly higher energy demand forecast, based on the assumption that world population will increase to over 10 billion, the WEC projects energy consumption growth to surpass 700 quads by 2050.[9]

Turning now to energy demand in the United States, as shown in figure 3.1, the EIA reported that total energy consumption in 2002 was approximately 98 quads, with 29 quads coming from imports, mostly oil and petroleum products. The EIA has projected that consumption will rise to 136 quads by 2025, with 53 quads coming from imports, an 82 percent increase in imports over the time period. Approximately two-thirds of the increase in consumption will be supplied by imported fuel. While total energy supply is projected to grow by about 1.5 percent annually, overall growth is projected to be about 40 percent over the period.[10]

In terms of energy supply, table 3.3 shows that while the EIA expects petroleum products, natural gas, and coal supply for the United States to grow by about 1.5 percent annually, it does not expect any noticeable growth in the supply of nuclear energy through 2025.[11] It can only be concluded that the EIA, unlike other divisions of the DOE, does not believe that any new U.S. nuclear plants will be brought into operation before 2025 under current policy. It is worth noting that hydropower supplied 75 percent of the renewable energy supply's contribution in 2002.

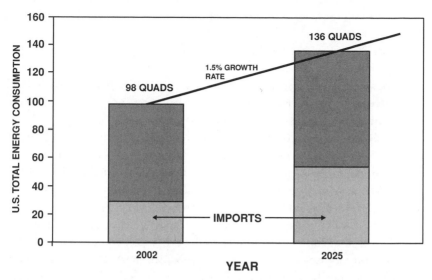

Figure 3.1.   Growing U.S. Energy Supply and Imports. *Source:* DOE/EIA, *Annual Energy Outlook 2004 with Projection to 2025*, DOE/EIA, January 2004.

Electricity Demand and Supply Outlook

World Electricity Outlook

Electricity plays an essential part in the daily aspects of life that we take for granted. For most of us, electricity is just there; it's invisible. We do not sufficiently appreciate that behind the wall socket is a huge complex of transmission lines, distribution stations, and power plants. These power plants can be fueled by coal, oil, natural gas, renewables, or uranium. Regardless of the energy source, each of us has experienced at one time or another how difficult everyday tasks become when our electricity is cut off

Table 3.3.   Breakdown of U.S. Energy Supply in Quads through 2025

Energy Source	2001	2025	Annual Growth
Petroleum products	38.1	55.0	1.6
Natural gas	23.4	32.2	1.4
Coal	22.2	31.7	1.6
Nuclear power	8.2	8.5	0.2
Renewables	5.8	9.0	1.9
Other (methanol, liquid hydrogen)	0.1	0.0	−4.6
Total	98	136	1.5

Source: DOE/EIA, *Annual Energy Outlook 2004 with Projection to 2025*, DOE/EIA, January 2004.

because of failures from faulty transmission lines or storm winds that knock down utility poles.

It stands to reason that while conservation can serve to reduce electricity demand in developed countries, those countries on the road to development will experience large surges in energy demand. Fuel-efficient technologies can undoubtedly help to moderate demand, but they cannot obviate the need for new power sources going forward in this century.

Furthermore, rapid urbanization of the industrializing countries will have an impact on the type of power needed. For example, densely populated cities will require baseload electric power supplies. In already industrialized countries, the demands of high-tech industries and communications networks require high-quality power sources that are reliably available—twenty-four hours a day, every day of the year. Unfortunately, renewables cannot provide electric power without nature-related interruptions. Therefore, not only must we consider how much power is needed, we must also provide power tailored to the needs of different sectors and populations. This translates into the need for baseload supplies of electric power.

Worldwide, 65 percent of electricity generation is produced from fossil fuels, with coal providing 40 percent, natural gas 15 percent, and oil 10 percent. Hydropower provides 18 percent of the world's electricity, nuclear fission provides 16 percent, and nonhydro renewables about 1 percent.

The fact that one-third of the world's population has no access to electricity, but will as countries develop, is one of the key reasons that electricity demand is projected to grow at 2.8 percent per year through 2010 and overall by 85 percent through 2020.[12] The EIA conservatively forecasts that worldwide net electricity consumption will double between today and 2025, from 13.3 billion kWh to 23.1 billion kWh. Growth in electricity demand is expected to increase in the developing world by 3.5 percent per year, and worldwide by 2.3 percent per year.[13]

The Massachusetts Institute of Technology (MIT) published a study in the summer of 2003, "The Future of Nuclear Power," which I will refer to frequently throughout this book because it adds yet another perspective on energy supply and demand from that offered by government institutions such as the EIA. This study forecasts that world electricity consumption will increase from 13,600 billion kWh in 2000 to 39,000 billion kWh by 2050, an almost tripling of use.[14] Worldwide, electricity consumption is projected to increase at an average annual rate of about 2.1 percent from 2001 to 2050.[15]

The Idaho National Engineering and Environmental Laboratory (INEEL) states that world electricity demand is rising at a rate of 2.7 percent per year and that 2,330 GW of new world electrical-generating capacity will be needed by 2020.[16]

Regionally, the most rapid projected growth in electricity use is for developing Asia, where robust economic growth is expected to increase demand for electricity to run newly purchased home appliances for air conditioning, refrigeration, cooking, and space and water heating. The EIA has projected that by 2025, developing Asia as a whole is expected to consume almost 2.5 times as much electricity as it did in 2001. In China, electricity consumption is projected to grow at an average of 4.3 percent per year by 2025, nearly tripling between now and then. For Central and South America, as in developing Asia, high rates of economic growth are also expected to improve standards of living and increase electricity use for industrial processes and in homes and businesses. Electricity consumption in the industrialized world is expected to grow at a more modest pace than in the developing world, with the EIA forecasting a 1.6 percent growth per year through 2025 and MIT projecting a 1.3 percent annual increase through 2050.[17]

U.S. Electricity Outlook

The EIA forecasts that the domestic generation of electrical power will have to grow from 3,443 billion kWh today to 5,257 billion kWh by 2025. This represents an overall growth of about 53 percent, or an annual growth rate of about 1.9 percent.[18] The EIA forecasts that the United States will need to add electric power station capacity equivalent to about 335 GWe (gigawatt-electric) capacity by 2025, even if ambitious assumptions are made regarding implementation of energy efficiency practices and technologies. Cumulative retirements in this time period amount to almost 62 GWe.[19] INEEL projects that in the U.S. alone there will be a need for 324 GWe of new capacity and 69 GWe of replacement capacity by 2020.[20]

It is expected that new electricity-generating requirements would result in an almost tripling of natural gas capacity, a 33 percent increase in coal-fired capacity, and a 20 percent increase in the so-called renewables capacity to 10 percent of total capacity by 2025. However, the forecast assumes no change in conventional hydropower capacity during the period. The EIA forecast assumes only a 4 percent overall increase in nuclear power capacity between now and 2025 as the result of power uprates and the anticipated return to operation of the Tennessee Valley Authority's Browns Ferry One reactor.

For the United States, while energy experts may differ as to the exact need for new electric capacity in the next fifteen years, the forecasts show there will be a requirement for between 324 and 342 1,000-MW electric power stations, or many more stations with a smaller capacity. Such an undertaking is large and may be regarded by some as mind-boggling and unlikely to actually happen. This points to the need for our nation to develop and put in place a national energy plan hard and fast before it is too late—if it is not already.

Electricity Generation by Oil, Natural Gas, Coal, and Renewables

The mix of primary fuels used to generate electricity has changed a great deal over the past three decades on a worldwide basis. Coal has remained the dominant fuel, although electricity generation from nuclear power increased rapidly from the 1970s through the mid-1980s. Natural gas-fired generation had grown rapidly in the 1980s and 1990s. In contrast, in conjunction with the world oil price shocks resulting from the OPEC oil embargo of 1973–1974 and the Iranian Revolution of 1979, the use of oil for electricity generation has been slowing since the mid-1970s. Continued increases in the use of natural gas for electricity generation are expected worldwide.

Oil

Petroleum, specifically oil, currently provides energy for the generation of 10 percent of the world's electricity, but 39 percent of the world's energy.[21] It is also used as a fuel for transportation and industry, as well as for both residential and commercial heating. In the United States at least 70 percent of oil consumed is used for transportation; additionally, oil supplies 40 percent of total U.S. energy demand. Most important, oil is also used to produce a long list of petrochemical products ranging from base chemicals to pigments to plastics.

During the economic "blahs" of the first few years of the twenty-first century, world oil demand grew approximately 1.8 percent. With the economic uptick now underway, the EIA projects that oil demand will start growing by about 1.9 percent annually. Our government's forecaster states that total world oil consumption will almost double, from 66 million barrels per day (MBPD) in 1990 to 121 MBPD by 2025.[22] Regions vary widely in their growth scenarios. In the developing Asian countries, oil demand should reach 3 percent annual growth rates while Europe faces a more subdued growth rate of 0.4 percent.[23]

The EIA forecasts that domestic crude oil production, which in 2002 was 5.6 MBPD, will decrease to 4.6 MBPD by 2025. Net imports are forecast to rise from 9.3 MBPD in 2001 to 15.5 MBPD by 2025.[24] In 2025, net petroleum imports, including both crude oil and refined products, are expected to account for 70 percent of demand, whereas imports in 2002 accounted for 54 percent.[25] Clearly we are heading in the wrong direction.

One subject that never fails to surprise me is the fact that many energy pundits give little attention to the imminent decline of oil and gas reserves during the next several decades.

Experts continuously argue over whether oil reserves will always be higher tomorrow than estimated today. Historically, resource base estimates have grown from 600 Gb (gigabarrels) in the 1940s to an estimated 3,900 Gb today. However, recently there have been questions as to whether the industry has reported up-to-date data, especially with respect to discovery versus use rates. Several companies have recently revised and lowered their estimates of economically recoverable resources. On January 9, 2004, Shell Oil Company announced a reduction in its proved reserves by 20 percent, or 3.9 billion barrels of oil equivalent. Specifically, 2.7 billion barrels of crude oil and natural gas liquids and 7.2 trillion cubic feet of natural gas were recategorized.[26] In February 2004, the El Paso Corporation reduced its proven oil and gas reserves by 41 percent, triggering a Securities and Exchange Commission investigation of the $1 billion reserve write-down.[27]

In its nineteenth edition of the "Survey of Energy Resources" in 2001, the WEC found that conventional commercial fossil fuels—encompassing coal, oil, and natural gas—remain in adequate supply, with a substantial resource base. Compared to its eighteenth edition of the same survey (published in 1998), oil reserve estimates actually declined slightly. In its commentary on oil, pessimistic and optimistic reserve assessments were compared and it appears that the former is more likely the case for the reasons quoted below:

- proved recoverable reserves of oil, which are largely concentrated in the Middle East, declined, while those of gas, which are more evenly spread, increased;
- fewer giant fields were discovered in the 1990s than in the 1960s (albeit a larger proportion were in deeper offshore waters);
- the discoveries of new oil fields were concentrated in a smaller number of countries in the 1990s than in the earlier periods;
- more recently the additional discoveries have been less than the oil produced;
- the oil industry's technological challenges posed by the ultra-deep offshore discoveries have not yet been met satisfactorily.[28]

At the present time we are only finding one barrel of oil for every four that we consume. Today about 6 Gb are discovered and 26 Gb are consumed. Most forecasts project that world production will peak by the end of this decade and start an inevitable long-term decline at between 3 and 6 percent. Middle East output is still rising but the rest of the world peaked in 1997.[29] Figure 3.2 conveys this picture.

Conventional methods of oil production provide most of the oil that we produce today, and have been responsible for 95 percent of the oil that has

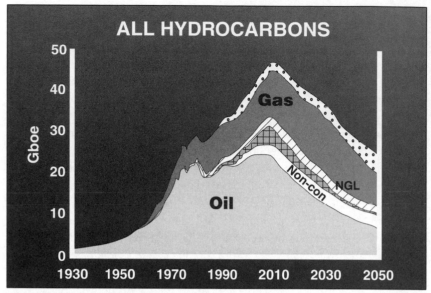

Figure 3.2.   Projected Peaking of Hydrocarbons (Gboe/year). *Source:* C. J. Campbell, online: www.hubbertpeak.com [accessed May 2004].

been produced in the world so far. It is expected that conventional oil production will start declining in the next year or so and that nonconventional production will delay overall peaking by a few years. It is estimated that production of all hydrocarbons will peak around 2010. Natural gas reserves, which are less depleted than oil, will likely peak around 2020.

We must accept the fact that oil supplies are finite, even if estimated reserves are temporarily ratcheted up. We already know the pressures put on our foreign policy due to the geographic location of oil supplies. Over 65 percent of the world's oil reserves are concentrated in the Middle East. It would be domestic energy policy suicide if we did not start planning today to switch to other fuels for home heating and transportation in order to save these precious resources for uses, such as for chemical feedstocks, that cannot handle substitutions. The day of reckoning regarding America's and the world's oil use is already on the horizon. The time to start preparing for it is now.

The bipartisan Energy Policy Act that I champion contains several measures to reduce our growing dependence on Middle East oil supplies and would move us to the next generation of fuel for transportation. Several of its key provisions would immediately have a positive impact on our oil crisis. The act proposed:

- $200 million for an advanced vehicle program to provide grants to state and local governments to acquire alternative fueled and fuel-cell vehicles, hybrids, and other vehicles, and another $200 million in grants for replacement of older school buses with clean, alternative-fueled and ultralow sulfur-fueled buses;

- increased funding for improvements in the fuel economy standards set by the federal government;
- a state-of-the-art program to get hydrogen-powered automobiles on the road by 2020;
- programs for hydrogen fuel-cell transit and school buses; and
- more oil and natural gas exploration and development by providing royalty relief for deep and ultradeep gas wells in the shallow waters of the Gulf of Mexico.

As a nation, we must face several realities. There are not enough votes in Congress to open the Arctic National Wildlife Refuge (ANWR) or the Outer Continental Shelf (OCS). Congress does not appear to have the stomach for sharply increasing Corporate Average Fuel Economy (CAFE) standards, mandating smaller cars, or imposing federal limitations on driving. Americans would never tolerate restrictions on the days one would be allowed to drive a car! Yet without some change in policy, our dependence on foreign oil is going to increase to 70 percent, up from 54 percent today. What we must do, then, is start working now to help shape a world oil market that is transparent, efficient, and fair. The United States must engage as many oil-producing countries as possible in the global oil market and strengthen our relations with non-OPEC producers. We must work on replacing oil used for transportation with hydrogen. And we must further diversify our sources of energy.

Natural Gas

Worldwide, natural gas use for electricity generation is projected to increase overall by 62 percent from 2001 to 2005, as technologies for natural gas-fired generation continue to improve and existing gas reserves are exploited. As shown in table 3.2, gas consumption is forecasted to increase from 93 quads in 2001 to 156.5 quads by 2025, an annual increase of 2.2 percent. The most up-to-date figures available predict that in the developed world natural gas will hold a 30 percent share of the electricity generation market and only 17 percent in the developing countries.[30]

Electricity markets of the future are expected to depend increasingly on natural gas-fired generation. Industrialized nations are intent upon using combined-cycle gas turbines, which usually are cheaper to construct and more efficient to operate than other fossil fuel-fired plants. Natural gas is much cleaner than other fossil fuels. In the developing world, natural gas is expected to be used to diversify electricity fuel sources, particularly in Central and

South America, where heavy reliance on hydroelectric power has led to shortages and blackouts during periods of severe drought.

The republics of the former Soviet Union (FSU) accounted for more than one-third of natural gas usage for electricity generation worldwide in 2001, and natural gas provided 42 percent of the energy used for electricity generation in the FSU sector. By 2025, natural gas is projected to account for 63 percent of the electricity generation market in the FSU sector. Relying increasingly on imports from Russia, the nations of Eastern Europe are also expected to increase their use of natural gas for electricity generation, from a 9 percent share of total generation in 2001 to 50 percent in 2025.

According to the EIA, natural gas consumption for electricity generation in Western Europe is projected to increase from 14.6 trillion cubic feet in 2000 to 23.7 trillion cubic feet by 2025, for an annual increase of 2 percent.[31] The natural gas share of the region's electricity fuel market is projected to grow from 17 percent in 2001 to 38 percent as the nuclear power and coal shares are reduced. After the oil crisis of 1973, European nations actively discouraged the use of natural gas for electricity generation (as did the United States), and instead favored domestic coal and nuclear power over dependence on natural gas imports. In 1975, a European Union (EU) directive restricted the use of natural gas in new power plants. The natural gas share of the electricity market in Western Europe fell from 9 percent in 1977 to 5 percent in 1981, where it remained for most of the 1980s. In the early 1990s, the growing availability of reserves from the North Sea and increased imports from Russia and North Africa lessened concerns about gas supply in the region, and the EU directive was repealed.

Per capita consumption of natural gas in Asia and Africa is relatively small when compared with Europe and North America. In 2001, Japan accounted for one-fourth of natural gas consumption in Asia. Almost all natural gas consumed in Japan is imported as liquefied natural gas (LNG). Japan is expected to maintain its dependence on natural gas at around 20 percent of the electricity fuel market through 2025.

In North America, the natural gas share of the electricity fuel market in the United States is projected to increase from 18 percent in 2001 to 24 percent in 2025, with Canadian exports expected to provide a growing supply of natural gas to U.S. generators. The natural gas share of electricity generation in Canada is also projected to grow, from 3 percent in 2001 to 11 percent in 2025.

In recent years, virtually all the electric plants (roughly 141 GWe of capacity) built in America have been powered by natural gas because it was thought to be cheap and abundant. It's turning out to be neither. Natural gas prices have more than doubled since 1999. In June 2000, contracts for gas to be delivered in July were $2.55 per million Btu. In June 2003, those contracts sold for $6.31, representing an increase of more than 150 percent. At the same time, gas reserve estimates are being revised downward. Canada has just reduced its natural gas reserves.

The combination of higher natural gas prices, lower LNG production costs, rising gas import demand, and the desire of gas producers to monetize their reserves is setting the stage for increased LNG trade. However, I question how this will come to pass with the opposition to siting the needed new LNG facilities. In the United States there are already four seaboard facilities for importing LNG and there are nine more in construction or planned. However, construction of LNG terminals in the United States is not without some controversy. There is opposition to plans in Eureka, California; La Joya, California; Mobile, Alabama; Vallejo, California; Searsport, Maine; and Fall River, Massachusetts. The residents of Harpswell, Maine, overwhelmingly voted in February 2004 to reject a proposed terminal. This foreshadows more battles and delays to come.

The continental United States imported 4.8 million MT (metric tons) of LNG in 2002, accounting for 4 percent of world LNG trade. The EIA reports that LNG imports were expected to approximately double in 2003 and increase tenfold to 46 million MT by 2010. The EIA projects that LNG imports from overseas will climb to about 100 million MT by 2025, twenty-four times the 2002 level. Global LNG liquefaction capacity is forecast to increase from 139 million MT in 2003 to 197 million MT by 2007, based on facilities currently under construction. While there were 151 LNG tankers operating in late 2003, fifty more were under construction, and more were needed. Three countries currently hold approximately 55 percent of the world's gas reserves: Russia (31 percent), Iran (15 percent), and Qatar (9 percent). The U.S. reserves amount to only about 3 percent of the world's total reserves, and hence our need for costly LNG imports—unless we look to alternative energy sources, like nuclear power.[32]

I am very concerned about the current and forecasted increases in U.S. demand for natural gas and what the resulting high prices mean for America's economy. We have lost more than two million manufacturing jobs in the last three years in large part because of soaring natural gas prices. As the chairman of the Federal Reserve, Alan Greenspan, testified before the Joint Economic Committee of Congress in May 2003, prices for natural gas have been rising because of very tight supplies:

> Working gas in storage is presently at extremely low levels, and the normal seasonal rebuilding of these inventories seems to be behind the typical schedule. The colder-than-average winter played a role in producing today's tight supply situation as did the inability of heightened gas well drilling to significantly augment

net marketed production. *Canada, our major source of gas imports, has little room to expand shipments to the United States. Our limited capacity to import liquefied natural gas effectively restricts our access to the world's abundant supplies of natural gas.* The current tight domestic natural gas market reflects the increases in demand over the past two decades. That demand has been spurred by myriad new uses for natural gas in industry and by the increased use of natural gas as a clean-burning source of electric power.[33]

This is déjà vu. Signs are brewing of a potential gas supply crisis similar to the oil crisis of the 1970s. More than 90 percent of the electricity generation built in America since 1996 relies solely on natural gas. As the oil crisis of the 1970s showed Americans, overreliance on one energy source got us into trouble. Today, with the natural gas situation, foreign supply is tightening, there is competition worldwide for the same resources, and our infrastructure is not up to par to bring in more imports, even if available. We are jumping from one frying pan into another!

The paths to increasing supplies of natural gas to U.S. markets are not environmentally or politically easy, fast, or cheap. We would have to immediately develop the Alaskan pipeline to bring in more supplies from the Arctic Circle and drastically increase LNG imports into the United States, which by the way could not happen until we add to the current four LNG ports.

I do not need to remind American consumers how hard rising natural gas prices could hit them. Natural gas prices have more than doubled since the late 1990s. Prices throughout 2003 were 60 to 65 percent higher than in 2002, and the trend continues into 2004. This directly affects the family pocketbook and it forces high-paying American jobs overseas. There is no question that the U.S. fertilizer business has been hit hard in recent years by high gas prices, and all reports point to the sad fact that the production of fertilizer components, urea and ammonia, will be shifted totally offshore.

The stakes are high. There is the real possibility that unprecedented increases in energy costs could seriously undermine the health of our economy for a long period of time. Increased supplies will be constrained in the near future by lack of infrastructure either in terms of pipelines or LNG ports and there is relatively little that can be done in the immediate short term to rectify the situation. But that does not mean that we should give up and bury our heads in the proverbial sands.

We simply should not be using natural gas for large-scale, centralized baseload power generation. Yes, it is suited to make peak power or for distributed generation or cogeneration applications. Gas does these things well: chemical feedstock applications, peak power, distributed generation, heating, and cogeneration. But it is a tragic waste of a precious resource to use gas for baseload electricity generation—especially when other forms of energy such as clean coal or nuclear power are better suited to the task.

Coal

Total world consumption of coal, 64 percent of which is used for electricity production, is projected to increase from 96 quads in 2001 to 140 quads by 2025.[34] Coal's share of world electricity generation was 34 percent in 2001, 40 percent by 2004, but is only expected to account for 31 percent of the world's electricity fuel market in 2005.

In Western Europe, coal accounted for 20 percent of the electricity fuel market in 2001 but is projected to have only a 12 percent share in 2025. The United States accounted for 40 percent of all coal use for electricity generation in 2001, and China and India together accounted for 27 percent. China's coal share is projected to rise only slightly, to 73 percent in 2025 from 72 percent in 2001. Over the same period, coal's share of India's electricity market is expected to decline from 72 percent to 63 percent.

The coal share of U.S. electricity generation is expected to remain at roughly 50 percent through 2025.[35] Coal's share of the electricity market peaked at around 57 percent in the mid-1980s.[36] There are some published reports, however, that coal is making more of a comeback. According to a DOE report by the National Energy Technology Laboratory, over the past three years there has been a huge jump in proposed coal-fired plants. It estimates that at least 94 plants are planned for an addition of 62 GWe.[37] The longer-term picture for coal use in the United States is colored by, on the one hand, the plentiful domestic supplies, and on the other hand, concerns over the volatility of natural gas prices as well as declining estimates of natural gas and oil reserves. Environmental concerns, some of which are talked about in this section, raise serious concerns about the long-term use of coal for electricity in the United States. Clearly there is a need for clean coal technology if we are to utilize coal for energy. However, with the cost of clean coal technology at one and a half times that of conventional coal fired plants, the market may not allow its introduction in the near future. Nuclear technology, available now, meets environmental goals for energy production without greenhouse gas emissions.

Reliance on coal for electricity generation is expected to decline in other regions. In Eastern Europe and the FSU, coal's 27 percent share of the electricity fuel market in 2001 is projected to fall to 6 percent in 2025. For years, massive state subsidies were all that kept many coal mines operating all over Europe. In many cases, the subsidies were underwritten by electricity consumers. The EU has adopted policy measures to eliminate or reduce state subsidies for domestic coal production, and only three of the original EU member states (the United Kingdom, Germany, and Spain) continue to produce hard coal.

The transport and handling of coal for a large steam-powered plant is a tremendous undertaking compared to that of uranium for a nuclear power plant. For perspective, take for example the Associated Electric Cooperative's New Madrid 1,200 MWe coal-fired electric power generating station, which is located on the Mississippi River just five miles from the city of New Madrid, Missouri.[38] The station's two 600 MWe generating units can each burn 8,183 tons of coal per day if fully utilized, or about 4.2 million tons per year, assuming a 70 percent capacity factor. This corresponds to about 11,500 tons per day. This coal is shipped 1,235 miles from Wyoming across four states, in unit trains composed of 115 lightweight aluminum rail cars containing 121 tons of coal each; each unit train hauls about 13,900 tons of coal, with a delivery every 1.2 days.

The magnitude of the environmental impact of hauling 4.2 million tons of coal annually may be compared to the hauling of about three hundred tons of natural uranium annually to fuel a 1,200 MWe nuclear plant. This equates to approximately 14,000 times less tonnage for production of the same amount of electricity. Compare also the annual emissions of carbon dioxide—little or none in the case of the nuclear plant and 8.4 million tons in 2000 in the case of the coal-fired New Madrid station.[39] Consider also the other pollutants emitted by the New Madrid station in 2000,[40] as follows:

nitrous oxide	35,861 tons
sulfur dioxide	16,451 tons
mercury	280 pounds
carbon dioxide	8,415,549 tons

The inescapable fact is that coal-fired plants now spew out high amounts of greenhouse gasses as well as other highly toxic materials. For example, there are radioactive elements in coal and fly ash. Studies conducted in the United States from 1975 to 1985 concluded that the maximum radiation dose to an individual living within one kilometer of a modern coal-fired power plant is equivalent to a 1 to 5 percent increase above the natural background radiation level. (In contrast, the typical increase from living at the boundary of a nuclear plant is about 0.3 to 1.0 percent above the average natural background level.)[41] While not currently regulated for radiation exposure, there is the potential for EPA regulations to clamp down on the radiation emissions from coal-fired plants in the future.

Mercury, on the other hand, is a larger issue for coal-fired plants. It is a public health issue that demands attention. The United States' mercury emissions total about fifty tons per year, most of which are emitted by coal-fired electricity generating plants. These plants remain the only major unregulated source of mercury emissions in the United States. As a result, there is a national debate over the need to control mercury emissions from such fuel-fired power plants.

Mercury is a toxic, persistent, bioaccumulative pollutant that affects human neurological development. Most exposure to mercury occurs through airborne mercury emissions and consumption of fish from lakes, ponds, and streams that have been affected by mercury precipitation from the air. Children exposed to mercury through their own (or their mothers') fish consumption are particularly susceptible to neurological damage. The National Academy of Sciences estimates that more than sixty thousand U.S. babies born each year are at risk for neurodevelopmental effects from mercury exposure. Many states currently have fish advisories to limit the consumption of fresh water fish due to high mercury concentrations.

A Clinton administration regulatory determination and court order required the EPA to adopt a "maximum achievable control technology" (MACT) standard for mercury that would require every power plant that exceeds the standard to install expensive emissions-control equipment or face penalties. This type of a MACT rule, depending on its stringency and implementation schedule, could be very costly (some estimate as much as $6 billion a year). Many proponents of mercury reductions agree that a 90 percent reduction (to about five tons per year from power plants) would be reasonable. But the industry has stressed that the technology is still developing and is prohibitively expensive. Other reports, including one by the Northeast States for Coordinated Air Use Management, found that mercury emissions from power plants could be reduced by 90 percent with regulation and commercially available technology.

As a compromise, President Bush proposed a smaller reduction of mercury pollution over a longer period of time. The EPA proposed rules in early 2004 based on regulatory options that include a 29 percent reduction by 2007 and 69 percent reduction in mercury emissions by 2018. These proposed reductions would allow trading of mercury-reduction credits, which has drawn extensive criticism due to the potential for high regional concentrations of mercury, unlike other pollutants that have been reduced through trading programs.

Will the utilities be able to meet stringent mercury standards and what will the cost be to the American consumer? Capturing mercury in coal-fired plants will be difficult. It is present in flue gas at concentrations of about 1 ppb (part per billion). For a comparison, picture the Houston Astrodome: the mercury in the dome would be equivalent to 30 out of 30 billion ping-pong balls. Capturing 90 percent of these ping-pong balls, 27 out of 30, as hoped for by environmental groups, would require you to find and remove 27 ping-pong balls out of 30 billion in the dome. Difficult at best and likely costly to the consumer![42]

Hydropower and Other Renewables

Renewable energy-generating technologies refer to conventional hydropower, geothermal, municipal solid waste, wood and other biomass sources, solar thermal and photovoltaic sources, and wind power. In 1990, total world consumption of hydroelectricity and other renewable energy slightly exceeded 26 quads (as compared to total energy consumption of 348 quads).[43] The EIA's most recent projection forecasts that world renewable energy consumption will grow at an annual rate of 1.9 percent, reaching consumption of 50 quads by 2025. In Europe, the United Kingdom and the Netherlands have significant plans to add renewable capacity, at annual growth rates of 5.5 percent and 6.2 percent, respectively. Among developing countries, South Korea tops the list with a forecasted annual increase of 5.9 percent.[44]

Renewables currently generate about 9 percent of U.S. electricity, with hydropower contributing 7.4 percent of U.S. electricity—making hydropower far and away the largest renewable source of electricity.[45] Hydro's role will not change appreciably during the next twenty years. Approximately 4.3 GWe of wind power are installed in the United States today, generating 10 billion kWh of electricity, enough to service about 1 million U.S. households.[46]

To put these energy sources into perspective, in the United States in 2002, total renewable electricity generation was 309 billion kWh,[47] whereas all other electricity generation sources provided 3,626 billion kWh to the grid.[48]

Renewable power generation in the United States is forecast by the EIA to grow at an annual rate of 1.9 percent per year to 464 billion kWh by 2025. While hydro will continue to provide the bulk of the electricity, no new hydro projects are on the horizon, and further, many existing dams are experiencing silting problems, reducing their capacity. The other renewable fuel sources are expected to show strong rates of growth, albeit from very small current levels. Comparisons of growth rates show annual increase rates of 0.8 percent for hydro, 5.6 percent for geothermal, 1.5 percent for municipal waste, 5.4 percent for wood and other biomass, 3.2 percent for solar thermal, 28.8 percent for solar photovoltaic, and 7.3 percent for wind power.[49] The high solar growth rates are not very meaningful since the starting base is very small.

Table 3.4 presents the EIA's analysis of U.S. electricity generation by renewable sources of energy.

One thing is clear: renewable energy facilities such as wind farms are running into environmental objections in the United States and elsewhere in the

Table 3.4.   Renewable Fuel Electricity Generation

Source	Generation (billion kWh)	
	2002	2025
Hydropower	255.8	304.8
Geothermal	13.4	46.7
Municipal solid waste	20.0	28.5
Wood and biomass	8.7	29.2
Solar thermal	0.5	1.1
Solar photovoltaic	0.0	1.0
Wind	10.5	53.2
Total	308.9	464.4

Source: DOE/EIA, Annual Energy Outlook 2004 with Projection to 2025, DOE/EIA, January 2004, table A17.

world, just as nuclear did. There is a range of objections to renewables. Hydropower damages river habitats and forces culturally and politically unpopular population resettlements. Photovoltaic solar panels require toxic chemicals for their manufacturing and large land areas. Geothermal sites are often located in pristine, protected wilderness areas that may need to be protected from development, give off some carbon dioxide emissions, require substantial water cooling, emit hydrogen sulfide, require the disposal of solids, and can be a depleting resource if not properly managed. Biomass power can lead to deforestation and soil erosion as well as emissions of carbon dioxide, nitrogen oxides, and particulates. Wind power is noisy, land intensive, construction materials intensive, and leads to significant avian and bat mortality. Interestingly, wind power supporters claim that wind turbines are less hazardous to birds than high-rise buildings, cars, transmission lines, and pet cats.[50] Central station thermal solar facilities have large land requirements and massive panel heights.[51]

Challenges to wind farms seem to know no bounds. A recent journalists' analysis of the status of wind farms in Europe concluded:

Voters are outraged by the unsightly turbines, the loud, low-frequency humming noise that they create and the stroboscopic effects of blades rotating in sunshine. Opponents are dismayed at the proliferation of the turbines in some of the most beautiful areas of the continent. Conservationists complain that hundreds of birds are killed each month by the rotating blades.[52]

Cape Wind Associates' plan to build 130 windmills in the Nantucket Sound has been met with a firestorm of protest over the threats to the sailing community, birds, and pristine ocean view. One of Scotland's most powerful environmental bodies, the Royal Society for the Protection of Birds, is opposing plans to site Europe's largest wind farm on the Isle of Lewis. The group argues that the proposed three-hundred-turbine wind farm would have a dramatically negative effect on protected bird species in the area such as the golden plover and golden eagles.[53] Opposition is surfacing in France, Denmark (the world's leader in wind technology), and the Netherlands. While Germany's parliament has approved a plan to double wind farms in the next sixteen years, the public is angrily protesting.[54] It's not only birds that raise environmentalist's ire, bats are being killed in droves at the Mountaineer Wind Energy Center on Backbone Mountain in West Virginia. Because of the large number of bats killed there from August to October 2003, the plans of the wind farm owner to construct many more wind turbines in that state and neighboring Pennsylvania and Maryland are threatened and may be delayed.[55]

Renewable energy sources are not expected to compete economically with fossil fuels in the mid-term forecast. In the absence of significant government policies aimed at reducing the impacts of carbon-emitting energy sources on the environment, it will be difficult to extend the use of renewables on a large scale. The EIA's 2003 projection indicated that consumption of renewable energy worldwide will grow by 56 percent, from 32 quadrillion Btu in 2001 to 50 quadrillion Btu in 2025.[56]

Much of the projected growth in renewable generation is expected to result from the completion of large hydroelectric facilities in developing countries, particularly in developing Asia, where the need to expand electricity production often outweighs concerns about environmental impacts and the relocation of populations to make way for large dams and reservoirs. China, India, Malaysia, and Vietnam, among others, are constructing or planning new, large-scale hydroelectric facilities.

Many nations of Central and South America have well-established hydroelectric resources. The nations of Central and South America are not expected to expand hydroelectric resources dramatically, but instead are expected to invest in other sources of electricity—particularly natural gas-fired capacity—that will allow them to diversify and reduce their reliance on hydropower. Hydroelectric capacity outside the developing world is not expected to grow substantially.

Among the other (nonhydroelectric) renewable energy sources, wind power has been the fastest growing in recent years. In Western Europe, Germany, Denmark, Spain, and other nations have installed significant amounts of new wind power capacity. Germany installed 2,659 MWe of new wind capacity in 2001, a national and world record for wind installation in a single year. In Spain and Denmark, wind power is doing so well that the governments are considering the elimination of subsidies aimed at promoting its installation.

The EIA's projections for hydroelectricity and other renewable energy resources include only on-grid renewables. Noncommercial fuels from plant and animal sources are an important source of energy, particularly in the developing world. The IEA has estimated that some 2.4 billion people in developing countries depend on traditional biomass for heating and cooking. However, comprehensive data on the use of noncommercial fuels and dispersed renewables (renewable energy consumed on the site of its production, such as solar panels used to heat water) are not included in the projections, because there are few comprehensive and reliable sources of international data on their use.

Some idea of the impact to our ecosystem of the alternate renewables sources of electricity generation may be gleaned from table 3.5. It is clear that the land impact of the renewables far exceeds that of nuclear power, by a factor of 120 or more for the photovoltaic and wind sources and enormous amounts in the case of the biosources.

Taking a wind farm for example, the American Wind Energy Association estimates that in an open, flat terrain, a utility-scale wind farm would require 50 acres per installed MW.[57] While proponents of wind farms argue that only 5 percent of the area is occupied by the turbines and the remainder can be used for other purposes, such as farming, a nuclear power plant can produce 1,000 MWe on 213 acres of land. In other words, we receive one thousand times the amount of electricity from a nuclear power plant on just four times the amount of land used for a wind farm.

While the debate over the competitiveness of renewables continues, I believe that the costs will come down and that we must develop renewables for energy security's sake. In 1978, I ran an ad favoring windmills in Clayton, New Mexico, and supported the development of solar panels at the Sandia laboratories. I was the only Republican senator who supported the Synthetic Fuels Act (S.1377) in 1977, demonstrating that throughout my career I have supported innovative approaches and alternative forms of energy with one goal—the enhancement of national and economic security of the United States.

Table 3.5. Comparison of Ecosystem Impact for 1,000 MWe Generation

Method	Requirement	Land Area (sq. miles)
Photovoltaic	100 km² @ 10% efficiency	40
Wind	3,000 wind turbines	40
Biogas	60 million pigs	>60
Bioalcohol	16,100 km² of corn	6,200
Bio-oil	24,000 km² of rapeseed	9,000
Biomass	30,000 of wood	12,000
Nuclear power	1 km²	0.33

Source: Dr. John M. Ryskamp, INEEL, IEEE Power Engineering Society Meeting, April 28, 2003.

The Energy Policy Act I support contains key programs and incentives for renewables. Highlights of its provisions include:

- reauthorization of the "Renewable Energy Production Incentive" program for solar, wind, geothermal, biomass, and other renewables;
- authorization of $300 million for solar programs with the goal of installing 20,000 solar rooftop systems in federal buildings by 2010 and $210 million for concentrating solar power for hydrogen production;
- grants of $550 million for biomass;
- authorization of $100 million to increase hydropower electricity production;
- direction to the federal government to use more renewable energy with the goal of using 7.5 percent more by 2011;
- provision of royalty relief for geothermal uses including on-site electricity generation;
- creation of substantial tax credits for many renewables; and
- requirement to add 5 billion gallons per year of ethanol and other renewable-based fuels to the nation's supply of gasoline.

I personally believe that we have reached the turning point where renewables are poised to make a small but significant contribution to diversifying our energy supplies. However, the intermittent and diffuse nature of several of the renewables do not suit them to provide reliable, 24/7 baseload electric power supplies in the current absence of economic energy-storage systems.

4

Nuclear Power in the World Today

More than fifty years ago a U.S. navy nuclear reactor illuminated a string of electric light bulbs to demonstrate that nuclear fission had potential civilian application. It has been fifty years since the first nuclear power plant was connected to the civilian electric grid in Obninsk, Russia, a city that I visited in 1998 when I gave a talk to a gathering of young students. The first U.S. civilian nuclear power generating plant, Shippingport, was brought into full power operation in December 1957 by the Duquesne Power and Light Company.[1] This event represented the beginning of the fulfillment of President Eisenhower's "Atoms for Peace" promise.

Between the mid-1950s and 1974, several hundred nuclear plants were committed in the United States, but unfortunately, less than half of them were actually built. Part of the reason for these failed plans was related to the obstructionist efforts of the antinuclear community whose tactics caused costly construction delays. The industry itself was also at fault because of its failure to standardize plant designs, and as a result, it stretched itself too thin and in too many directions, and at a cost.

I believe that the nuclear industry, at least in the United States, has learned much from the past and is now perhaps the most efficient and safe energy industry in the world, a fact that its economics and vital statistics support. This is a message that I strive to get out to the public.

Today there is as much electricity generated by nuclear power alone in the United States as there was generated by all sources worldwide in 1960. There are currently 103 operating nuclear plants in the United States with a total capacity of 98 GWe, providing 20 percent of the nation's electricity supply. There is one additional plant licensed in the United States that is operable but that has been shut down for some years—Tennessee Valley Authority's Browns Ferry Unit One. It is scheduled to restart in 2007, pending final NRC approval.

Worldwide there are 437 operable nuclear plants in thirty countries with a total generating capacity of 362.9 GWe, supplying approximately 16 percent of the world's electricity. There are thirty plants under construction and thirty-three ordered or planned.[2] Table 4.1 lists those countries with significant nuclear power programs as of December 2003. Europe obtains 35 percent of its electricity from nuclear power, thanks to the lead taken by France in the 1970s, making long-term use of nuclear power a key factor in that continent's clean air strategies and obligations, as well as electricity dependence. Unfortunately, misguided activists have put political pressures on some governments to phase out their nuclear programs in the coming years, most notably, those of Germany, Belgium, and Sweden. I hope that the public and industry in these countries can persuade their governments to see the folly of their policies and cancel their nuclear phase-out plans. Without nuclear power, these countries will not meet their reduced pollution emissions obligations and will also suffer severe negative economic impacts.

The United States and France together account for about 45 percent of the world's 363 GWe of nuclear power capacity. If Japan, Germany, and Russia are included, five countries currently make up two-thirds of the world's capacity.

As table 4.1 shows, nuclear plant capacity factors range from 71 percent for Russia to 91 percent for the United States and South Korea. The Institute of Nuclear Power Operations (INPO) released its 2002 performance data for U.S. plants, showing that for the third straight year, the industry's unit capability factor topped 90 percent, reaching a record high 91.2 percent. The average nuclear capacity factor for the United States was down slightly in 2003 because of the extended outage of the Davis Besse 1 nuclear plant. I should point out that the 75 percent capacity factor for France should not be taken as a reflection of that country's operating efficiency, but is due instead to the

Table 4.1. Countries with Significant Nuclear Power Programs

Country	Nuclear Power Capacity (GWe)	Number Plants	2002 Capacity Factor (%)	2002 Nuclear Share (%)
United States	97.3	103	91	20
France	63.2	59	75	78
Japan	44.0	53	81	34
Germany	21.3	19	87	30
Russia	20.7	26	71	16
Rep. of Korea	14.9	18	91	39
United Kingdom	12.3	31	74	22
Ukraine	11.2	13	75	46
Canada	10.0	14	84	13
Sweden	9.4	11	79	46

Source: Based on World Nuclear Association fact sheets.

fact that many French reactors are operated in what is called the "load-following mode"—that is, they go up and down with the daily requirement for power.

World Nuclear Power Forecasts

Published nuclear power forecasts range from pessimistic to extremely optimistic, and call for careful consideration from the "what is realistic?" point of view. Most of the forecasts address the near- to mid-term—that is, the period between now and 2020 to 2025. However, several institutional studies look at the longer term—that is, through the middle of the century and beyond. Differing underlying assumptions and forecast dates can cause understandable confusion in comparing the various projections.

The most notable of the longer-term forecasts is that projected in the International Institute for Applied Systems and World Energy Council (IIASA/WEC) 1998 study.[3] This study projects a world nuclear growth scenario to 2,000 GWe by 2050 and 6,000 GWe by the end of this century! The IIASA/WEC forecast assumes that demand will include the forecasted hydrogen economy based on nuclear power.

A study done at MIT in 2003 defines a "mid-century" scenario in which the generating capacity of nuclear plants in the world grows to a range of 1,000 to 1,500 GWe.[4] The 1,000 GWe forecast reflects a relatively constant nuclear electricity market share of about 18 percent between now and 2050 with electricity demand growing at a rate of about 2.0 percent annually. While the MIT forecasts may be regarded as optimistic, they are considerably less so than the IIASA/WEC forecast.

The MIT study group assumed that by 2050, the world's installed nuclear capacities would be distributed approximately as follows:

United States	300 GWe
Europe & Canada	210 GWe
East Asia	115 GWe
Former Soviet Union	50 GWe
Developing World	325 GWe
Total World	1,000 GWe

If the MIT forecast comes to pass, growth in nuclear power in other parts of the world will be equally strong as compared to the United States, taking more than twice the U.S. effort to meet global demand. It is difficult to envision how the world could meet the IIASA/WEC forecast other than by many countries mounting efforts analogous to the U.S. and Russian space races in the 1950s and 1960s.

The International Atomic Energy Agency (IAEA) Nuclear Consultancy Capacity group's July 2003 low and high forecasts of world installed nuclear power for 2010, 2020, and 2030 are presented in table 4.2. The table also shows the IAEA's estimates of world total electric capacity for each of those years.[5]

Table 4.2.   IAEA Forecasts of World Nuclear and Total Electric Capacity (GWe)

	2002	2010	2020	2030
Nuclear power				
Low		391	423	386
Mean	359	396	462	480
High		401	501	573
Total electricity				
Low		3,791	4,378	5,064
Mean	3,469	4,019	4,956	6,167
High		4,246	5,534	7,270

Source: IAEA, Energy, Electricity and Nuclear Power Estimates for the Period up to 2030, IAEA, July 2003.

The EIA's outlook on the future of nuclear power in the world continues to be pessimistic, though a little less so than it was in previous years.[6] Under the EIA's "reference case," worldwide operable nuclear capacity, which was 358 GWe in 2002, is projected to plateau at 407 GWe in 2015, and then decline significantly to 385 GWe by 2025.[7] While the projection is a glaring improvement on EIA's 2002 nuclear power outlook, which showed capacity increasing to only 363 GWe by 2010 and declining to 359 GWe in 2020, it still does not seem to be in tune with reality.

The EIA attributes the increase in nuclear output to higher capacity utilization rates and the expectation that fewer existing plants will be retired. The EIA admits that on the optimistic side, for example, emerging technologies could change the economics and perceived safety of nuclear power

Table 4.3.   EIA Forecast of World Nuclear Capacity (GWe)

	2002	2010	2020	2025
Nuclear power				
Low		369	330	273
Reference	358	392	401	385
High		406	485	538

Source: DOE/EIA, Energy, International Energy Outlook 2004, DOE/EIA, April 2004, tables E1, E2, and E3.

plants, as well as public sentiment about radioactive waste disposal and nuclear weapons proliferation. In the EIA high nuclear forecast case, world nuclear capacity is projected to grow from 358 GWe in 2002 to 538 GWe in 2025. However, the EIA's pessimistic forecast is for only 273 GWe to be in operation by 2025.[8]

The IAEA "mean" forecast for nuclear power in 2020, which represents the mean between the low and the high forecasts, is considerably higher than the EIA reference forecast for the same year. The IAEA mean for 2025, based on interpolation of the 2020 and 2030 forecasts, is 471 GWe, more than 86 GWe greater than an EIA forecast.

Nuclear Power in the United States Today

Nuclear power is the second leading source of electricity generation in the United States, after coal. There are presently 104 licensed nuclear power plants in the United States, with a combined net capacity of 97,300 MW. This includes the Browns Ferry Unit 1 plant, a 1,065 MWe unit belonging to the Tennessee Valley Authority that has been out of service since 1985.

Nuclear power plants in the United States achieved record-high electricity output in 2002, producing 785 million MW hours of electricity, compared to 769 million MW in 2001 and 557 million MW in 1999. In 2003, these plants produced 762 million MW of electricity, enough to serve 75 million homes in America.[9] Nuclear power plants currently provide electricity to one in every five homes and businesses in the United States, representing 70 percent of the electricity from sources that do not pollute the environment.

Since 1990, there has been an increase in electricity output equivalent to the amount that would be generated by twenty-five new 1,000 MW nuclear plants. This has largely been the result of increasing capacity factors and power uprating. From 1998 until 2002, increases in nuclear power plant efficiency were equivalent to adding thirteen new 1,000 MW power plants to the U.S. electricity grid.

In spite of the relatively poor economy during the first few years of the twenty-first century, the low production costs, high capacity factors, and air-pollution-free operation of nuclear power plants are being positively recognized in the energy and financial industries as well as with policymakers. The number of nuclear plants that have announced plans to seek license renewal continues to increase.

Under NRC regulations, licensees are allowed to seek license extension for power reactors after twenty years of operation. The Atomic Energy Act and

NRC regulations limit commercial power reactor licenses to an initial forty years, but also permit such licenses to be renewed after twenty years. This original forty-year term for reactor licenses was based on economic and antitrust considerations—not on limitations of nuclear technology. The length of time between the filing of an application and the granting of a license has been approximately two years, which the utility industry regards as satisfactory.

The present status of U.S. nuclear plant license renewal is presented in figure 4.1. Through the year 2020, the original operating licenses for fifty-seven nuclear power plants, totaling 45,000 MW, were scheduled to expire. Since license renewals began in March 2000, twenty-three plants have been granted twenty-year license renewals by the NRC, extending their operating lives until between 2033 and 2043. The NRC is currently evaluating applications for the extension of operating licenses of fifteen more nuclear units, and many more applications are expected. The U.S. operators have notified the NRC of their intent to submit license renewal applications for at least sixteen additional nuclear power plants during the next few years. The NRC anticipates that 70 percent of all U.S. nuclear reactors will either have applied for or received extensions of their operating licenses within the next five years.[10]

License extension is expected to allow the industry to maintain at least the current level of generation over the next fifteen to twenty years—that is, un-

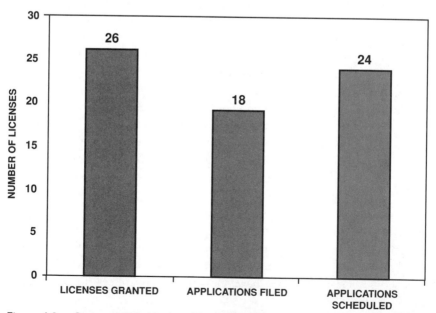

Figure 4.1. Status of U.S. Nuclear Plant License Renewals. *Source:* NEI and NRC data, May 2004.

til existing plants run their full lives and, hopefully, new plants are started up. These are not signs of a dying industry. *Nuclear power is on the comeback trail in America.*

The Outlook for Nuclear Power in the United States

The EIA forecasts a minimal 0.2 percent annual growth in nuclear power up until 2025 because it does not assume a new nuclear plant will be brought into operation by that time. Nuclear power generation is forecast to increase from 780 billion KWh in 2002 to 816 billion KWh in 2025.[11] The U.S. nuclear generation capacity forecast increases only slightly from 98.7 GWe in 2002 to 102.6 GWe by 2025.[12]

However, I am quite skeptical of the EIA's forecasting methodology. In 1999, a critique of the EIA's methodology for forecasting the future growth of nuclear power in the United States concluded that the economic analyses upon which the forecasts were based at the time did not accurately reflect market conditions.[13] The EIA uses the National Energy Modeling System to forecast the future growth of all energy forms except nuclear power, and then externally estimates nuclear power costs to determine whether or not nuclear power plants are competitive with new natural gas combined-cycle plants based on clearing price considerations. While this general approach was not necessarily incorrect, it required that the underlying nuclear power cost data be accurate and up-to-date, something that it was not several years ago.[14]

While the EIA has made some of the corrections recommended by its critics in 2000 and 2001, some forecasting shortcomings seem to still exist. It is of interest to note that the projected generation for the year 2020, which is the latest date common to all forecasts, has increased significantly in each *International Energy Outlook* report during the past five years, as shown in figure 4.2.

The EIA has begun to change its view of the nuclear option, and as a result, has during the past six years more than doubled the projected installed capacity for the end of the next decade. EIA is now willing to accept that fewer existing U.S. plants will be retired. However, it is still reluctant to believe that there could be any new plants brought online between today and 2025.[15]

Although I take nuclear power growth forecasts with a large grain of salt, respectable organizations such as our national laboratories and MIT envision the need for significant increases in the number of nuclear power plants in the United States. For example, the previously mentioned MIT study postulates that the United States may have up to 300 GWe by 2050, which would require the addition of about 5 GWe annually between 2012 and 2050, approximately four 1,300 MWe-class reactors each year. While this may be

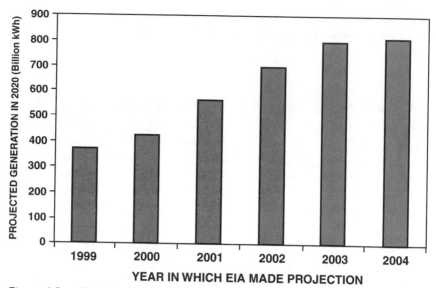

Figure 4.2. EIA Projections of U.S. Nuclear Power Consumption in 2020. *Source:* EIA *Annual Energy Outlook* reports.

doable, it will take a great change in the utility industry's and the investment community's current wariness toward nuclear power. However, I believe that it can and must be done if this country is to move forward while at the same time protecting the environment.

As shown in figure 4.3, the study performed by six DOE-sponsored national laboratories projects a U.S. nuclear power growth to about 500 GW by 2050, assuming development of Generation IV technologies and advanced light water reactor (LWR) nuclear-generated electricity.[16] The production of hydrogen using high-temperature advanced reactors could result in nuclear power capacity of about 730 GW by 2050, as seen in the figure. These are fivefold and sevenfold increases, respectively, over today's domestic nuclear generating capacity. This study postulates that nuclear power could provide 23 percent of U.S. electricity—about 200 GW by the year 2020 with power uprates and advanced light water reactors.

I am pleased that the DOE is helping to pave the way for new reactor orders through its Nuclear Power 2010 Initiative. DOE's goal is to enable the U.S. utility industry to deploy at least one new advanced nuclear power plant in the U.S. by 2010. This goal is based on the assumptions that there will be an economic, commercial-scale hydrogen production system using nuclear energy by 2015, and a next-generation nuclear system for deployment after 2010 (but before 2030) that provides significant improvements in proliferation and terrorism resistance, sustainability, safety and reliability, and economics.[17]

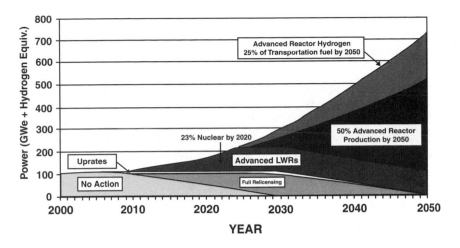

Figure 4.3. Recommendations for the Expanded Use of Nuclear Energy in the United States. *Source:* Argonne National Laboratory et al., *Nuclear Energy: Power for the Twenty-first Century*, Argonne National Laboratory Report ANL-03112, May 2003, figure 5.1.

Three nuclear plant design certifications already exist and a fourth (the AP-1000) is under review. Three utilities (Dominion, Entergy, and Exelon) have begun the early site permit process. In October 2003, Entergy and three other companies submitted an application to the NRC for permits to site new reactors at existing nuclear plant sites. Entergy's permit application is for the Grand Gulf site in Mississippi. Exelon and Dominion Energy filed permits for sites in Illinois and Virginia, respectively.

On the eve of the twenty-fifth anniversary of the Three Mile Island accident, in whose aftermath all pending reactor orders were cancelled, a significant step forward was taken by the nuclear industry. The formation of three consortia to pursue activities that could lead to the first new nuclear plant order in twenty-five years was announced. The first consortium comprises Dominion Energy, Hitachi America, Bechtel Corporation, and AECL Technology, whose reactor design is the Canadian Advanced Candu Reactor-700. The second group, called NUSTART, includes Exelon Generation, Entergy Nuclear, Constellation Generation, Duke Energy, Progress Energy, Florida Power & Light, Southern Company, TVA, EDF International North America, Westinghouse Electric, and GE Energy. These partners have collectively agreed to seek DOE funding and support to demonstrate and test a new process for obtaining the Combined Construction and Operating License (COL) for an advanced nuclear plant. Each member of that group has committed $1 million a year in cash plus in-kind and administrative services over a period of seven years. The group plans to complete a COL application and submit it to the NRC in 2008, hoping for a positive NRC decision in late 2010. It is expected that any combination of consortium members could join forces at that time to order and build a new nuclear plant. A third consortia consisting of TVA, Toshiba, USEC, Global Fuel–America, Bechtel Power, and General Electric submitted an application to DOE for a feasibility study to build an advanced BWR at TVA's Bellefonte site.[18]

The Nuclear Energy Institute (NEI) envisions the nuclear industry deploying more than 10,000 MW of additional power at existing plants in the next decade by increasing capacity factors, power uprates, restarts, and license extensions. The nuclear industry, through the NEI, has put forth an ambitious "Vision 2020" plan to bring 50,000 new megawatts of nuclear capacity online by that date. That would translate into fifty more nuclear power plants in the United States. If we cannot completely reach this goal, let's hope that we can at least head toward it.

In my judgment, we are witnessing a rekindling of utility interest in nuclear power in the United States. The outlook for nuclear power in the United States has not been as rosy since the late 1970s and early 1980s. This promising future can be attributed to the convergence of many, many factors:

- progress on extending Price Anderson indemnification for nuclear accidents;
- efforts to reduce licensing risks through "early site permits" and the COL process;
- the economics of advanced, passively safe nuclear plants are now competitive with the lowest cost alternatives, natural gas, and combined cycle;
- a desire in both the public and private sectors for fuel supply diversity to help manage market risk;
- safety improvements are reducing economic risks associated with plant ownership and operation;
- the prospect that other states may follow New Hampshire's lead in offering pollution emission credits to new nuclear plant capacity obtained through either new construction or capacity expansion—the Seabrook plant will receive nitrogen oxide (NO_x) emission credits when its planned 6 percent power uprating project is completed in 2005[19] (such credits will help to level the playing field);
- five straight years of increases in electricity production by nuclear plants worldwide;
- dramatic increases in nuclear plant capacity factors, from 60 percent in 1987 to 91 percent in 2002;
- production costs of 1.7 cents kWh that are slightly lower than those for coal and much lower than natural-gas fired plants at 3 to 5 cents;

- an excellent safety record unmatched by any other manufacturing industry in the United States;
- plant license extensions and power uprates;
- the shaping of a National Energy Policy that favors a balanced approach, including nuclear power in the energy mix;
- President George W. Bush's push for the development of a hydrogen economy that would require new nuclear baseload capacity;
- increased public awareness of the role nuclear power can play in the reduction of greenhouse gasses;
- a recent uptick in public support for nuclear energy;
- consolidation of nuclear power utilities into stronger, larger companies;
- promising designs for next-generation nuclear power plants that are more proliferation resistant and inherently safe; and
- increased momentum in obtaining a construction license for the first nuclear high-level waste repository.

Status of Major Nuclear Programs Outside the United States

Examining the status and prospects for nuclear power, we recognize that as few as six countries are expected to provide more than 70 percent of the world's installed nuclear capacity during the next ten to twenty years. The United States, France, and Japan will continue to be the dominant commercial nuclear generators through the year 2020. Japan, China, South Korea, and Russia could maintain the largest construction programs during the same period.

Taking a look around the world, we can see that nuclear power is holding its own in countries already wedded to it and is gaining favor in some other countries. Finland has committed to a Framatome-Siemens 1,600 MWe European pressurized water reactor (EPR) based on its clear economic advantages for electricity generation for Finnish industries. France is undertaking an official national energy policy debate and is expected to commit to building new plants. While the French government has announced its intentions of building a demonstration EPR plant, Finland's commitment has gone beyond demonstration to implementation.

While seven of the EU's fifteen member nations currently have operating nuclear power plants, three of these seven countries—Sweden, Germany, and Belgium—have unwisely developed policies to shut down their plants as soon as possible. Policy aside, however, because of the importance of nuclear-generated electricity plants in their energy mix, Germany and Sweden will continue to operate their reactors for their full licensed lifetime.

Thinking of Germany's situation reminds me of my trip to Europe in the summer of 1998. After visiting France and Russia, we stopped in Germany. My main purpose was to inspect the facility at Hanau that Germany had begun in 1987 for the purpose of providing MOX fuel to German reactors. Construction on the facility was stopped in 1991. At the time the plant was being proposed as a way of jump-starting MOX fuel production in Russia and moving faster with disposition of weapons-grade plutonium. During discussions with my hosts, the subject of the German Green Party's opposition to nuclear power came up. The plan in Germany, driven by the Green agenda, was to shut down all nuclear plants, despite the fact that they then provided about one-third of German electricity. I asked my hosts where Germany would get their power if those plants were shut down. I've never forgotten the sincere answer that Germany would simply buy the power from France. It struck me as amazing logic to, on the one hand, argue against nuclear power, and at the same time, accept that the only source of power remaining to Germany would be clean nuclear power from its next-door neighbor.

Nuclear matters are brighter in other countries such as France, Finland, South Korea, and China. The remainder of this chapter highlights major nuclear programs around the world. For an industry whose critics like to call dying, the fact that there are 337 nuclear plants with a combined capacity of 264 GWe operating outside of the United States is just one further indication of nuclear power's importance in the energy picture. Even though programs of the United States, France, and Japan dominate the worldwide picture of nuclear power, there are significant programs in a whole host of other countries.

France

France currently has fifty-nine operating nuclear power plants with a combined net capacity of 63 GWe. Over 75 percent of the country's domestic electricity demand is currently being met by nuclear power. I fully expect France to maintain its commitment to nuclear power. As part of a series of public sessions, Industry Minister Nicole Fontaine has argued that renewables and conservation cannot provide the energy France will need in the future and that the choice will come down to greater reliance on fossil fuels or nuclear power. Two parliamentarians, Claude Birraux and Christian Bataille, recommended the construction of an EPR without delay. At the opening of the third session of government under his watch, Prime Minister Jean-Pierre Raffarin declared that his government is committed to building a new nuclear power plant.

France exports large amounts of nuclear power-generated electricity to Switzerland, Italy, and the United Kingdom, with smaller exports to Belgium, Holland, Spain, and Germany. Competition in the European electricity market has led to an increase in electricity exports, and France continues to open its electricity market to foreign competition in accordance with EU rules. By the end of 2004, 70 percent of the French electricity market (industrial and commercial, but not residential) will be fully liberalized or open to competition. Italian utilities are exploring the possible purchase of some French capacity.

I have been particularly impressed by the success of the French program, which has taken advantage of nuclear plant design standardization and nuclear fuel-cycle closure almost from the outset. They have been able to do so because of the largely centralized government control of both their utility and fuel cycle industries. They have also taken advantage of opportunities to export their nuclear technology expertise. It is ironic that French LWR technology deployment began with the licensing of American taxpayer-funded pressurized water reactor technology provided by the Westinghouse company in the 1970s, at a time when domestic utilities were pursuing a wide variety of different reactor designs.

An example of the leadership of the French nuclear program is the recent award to its reactor vendor, Framatome ANP, of a contract to build the EPR in Finland. There are reports that Framatome ANP will file an application with the NRC for design certification in the United States. Prior to this award, an early order of the EPR by EDF had been considered essential if the current fleet of operating nuclear power plants were to be replaced in a gradual and orderly manner starting around 2020. A commitment to the EPR is expected in the near future, especially now that Finland has led the way forward.

It is likely that that French nuclear capacity will increase to approximately 66 GWe by 2020. This projection assumes that licenses will be extended for units reaching forty years of operation by 2020.

Japan

Japan presently has fifty-three nuclear power plants with 44 GWe of net capacity in operation, providing 34 percent of its electricity-generation needs. Construction of three additional units is well underway. Japanese utilities expect to raise annual average capacity factors from their current level of about 80 percent to better than 85 percent with the government approval for an increase of the legal operating cycle length from 13 to 16.5 months.

The Japanese nuclear industry has been hard-hit by scandals ranging from falsification of operating records to operating procedures resulting in accidents. Because of these incidents, the Japanese government has given up its ambitious goals of the late 1990s to increase nuclear generation capacity to 70

GWe by the year 2010. Nonetheless, the government announced in 2002 that a heavy reliance would have to be placed on nuclear energy to achieve green-house gas emission reduction goals set by the Kyoto Protocol. Despite the problems discussed above, policy leaders, including those suspicious of nuclear power, all seem to realize that the nuclear option is unavoidable for energy resource-poor Japan, particularly in light of goals that have been established under the Kyoto agreement regarding carbon emissions.

It is likely that installed Japanese nuclear capacity will rise to 48 GWe in the year 2010 and 57 GWe in 2020. This prediction assumes that Japan adds necessary new nuclear capacity through 2020.

The United Kingdom

The United Kingdom, one of the earliest nuclear power pioneers, has twenty-seven operating nuclear power plants with a combined net operating capacity of 12 GWe, providing 22 percent of the country's electricity generation. While no new nuclear units are currently under construction or firmly planned at this time, and wind power and other renewables are being em-braced, future deployment of nuclear power is still considered an option.

The British government completed the privatization of most nuclear gen-eration assets with the public offering of British Energy in 1996. In June 1998, the Nuclear Electric and Scottish Nuclear subsidiaries of British En-ergy, formed a single subsidiary British Energy Generation Limited. This company operates the United Kingdom's AGR (advanced gas-cooled reac-tor) and PWR nuclear power plants. Operation of the older Magnox stations was transferred to British Nuclear Fuels plc (BNFL). BNFL has a "lifetime strategy" whereby the Magnox units will be phased out between 2002 and 2010. The Calder Hall station was retired in March 2003 and the Chapel Cross station is expected to be fully retired by 2005. The larger Magnox units will have achieved forty-year lifetimes while the smaller units will have achieved fifty-year lifetimes. The retirement strategy is driven by the eco-nomics relating to the decision to shut down the Magnox fuel reprocessing line at Sellafield by 2012.

British Energy's attitude toward new nuclear investment turned positive in 2001, and the company considered the replacement of existing nuclear assets with new nuclear capacity upon their retirement. These plans were placed on standby when British Energy fell into severe financial crisis during 2002 due to low power pool prices. British Energy was ultimately forced to take a bailout loan from the government and to sell off its North American invest-ments in Bruce Power and Amergen in order to raise cash. A February 2003 government white paper by the Royal Academy of Engineering ruled out consideration of new nuclear power plant construction until at least 2006. The government instead decided to place all its energy policy emphasis on renewables and energy efficiency. The new policy calls for increasing renewables' share of electricity production from less than 3 percent currently to 10 percent by 2010 and 20 percent by 2020. Nuclear's role as an important source of carbon-free electricity was noted, but its current economics were judged to be unattractive for new, carbon-free generating capacity. The government policy paper does not seek early closure of currently operating units, and noted that the severe financial problems of British Energy during the past year were specific to that company and not indicative about the future of nuclear power. Not unexpectedly, the white paper has received heavy industry criticism, particularly in regard to what are considered unrealistic goals for renewable energy generation. New nuclear generation will be reconsidered if progress on the renewables goals appears slow, but this evaluation will not be made until the new energy policy has had five years to prove itself.

While it is possible that one or two new PWR plants could enter service in the year 2020, the current prospects for this are very uncertain. Twenty-four units of Magnox and AGR capacity totaling 7.5 GWe could be retired between now and 2020. As a result, total installed nuclear capacity could decline to between 6 and 8 GWe by 2020.

Germany

Germany currently has eighteen operating nuclear power plants with a combined net capacity of 20.6 GWe, providing about 30 percent of the country's total electricity generation. Unobstructed operation has been a problem for plants located in Social Democratic Party (SPD) governed states, as local licensing authorities are antinuclear. After years of negotiations, the SPD/Green Party coalition was able to strike a deal with German nuclear power plant operators regarding the remaining operating lives of Germany's nuclear power plants. The phase-out agreement was made law through an amendment to the Nuclear Power Act in 2001, limiting the remaining generation available to German plants. In exchange, the government will allow uninterrupted operations and waste management, provided safety standards are met. The phase-out agreement also calls for the end of reprocessing of spent nuclear fuel by July 2005, when transport to reprocessing plants will be halted. While characterized as capping German nuclear power plant lifetimes at thirty-two years, the phase-out agreement actually sets the remaining generation allowed by each unit, and allows unused generation at some units to be transferred to the generation limits of other units owned by the same company. The generation transfer feature is apparently open to further negotiation, as the transfer of generation rights to the Obrigheim plant has been limited to two years' worth by the government, rather than the five years proposed by the utility. Assuming future capacity factors of 85 percent, the average lifetime under the agreement is estimated at thirty-four years. Some of the more radical members of the SPD/Green coalition do not feel the agreement is severe enough on nuclear, but the government appears to be committed to the terms of the agreement. Some of the nuclear utilities have subsequently stated that they have not agreed to phase out nuclear power, only to limit the lifetimes of existing units, apparently leaving open the opportunity for new or replacement nuclear capacity.

Consistent with the phase-out agreement with the German government through 2010, two units are projected to be shut down by 2005 and another four between 2007 and 2010. The terms of the phase-out agreement are assumed to be eased after 2010, allowing remaining units to reach forty-year lifetimes, although no new capacity additions are allowed. The result is a decline in installed German nuclear capacity from 20.6 GWe in 2002 to 14 GWe by 2020, with the number of operating units declining from eighteen to ten. One can only hope that the government realizes the importance of nuclear power to Germany's economy before it is too late.

Sweden

Sweden currently has eleven operating nuclear power plants with a combined net capacity of 9.4 GWe. Although nuclear power provided almost half of the country's electrical generation during 2002, there are plans in place to eventually shut down all of the country's nuclear power plants in accord with a 1980 referendum. While the current nuclear phase-out policy seems to make no sense because of Swedish industrial requirements, it is likely to remain a political issue in Sweden for a number of years. Swedish public opinion appears to be changing and may be favorable to a restart of the national nuclear program. The principal opposition party in Sweden, the Liberal Party, strongly supports nuclear power as a way to meet the country's electricity needs while maintaining its emissions goals. Nuclear capacity is currently projected to remain at 9.4 GWe through the year 2020.

Sweden's waste management policy, like that of the United States' today, mandates the direct disposal of spent fuel. Sweden has operated a central spent-fuel storage facility since 1985, and has a final repository for low- to intermediate-level wastes now in operation. Sweden has an underground research laboratory for disposal of high-level waste, which is providing information for the ultimate repository, slated to begin accepting wastes in 2015. Site selection is now underway. Site investigations were initiated in Sweden in 2002, and the final decision is scheduled to be made no later than 2010. SKB reports it anticipates Sweden will be in a position to place the first canister of spent nuclear fuel in the deep repository by 2015.[20] In a 2003 survey of eight hundred people living near the two communities under consideration for the site of the repository, completed for the Swedish Nuclear Fuel Waste Management Company, 67 percent in one location and 65 percent in the other favored construction of the facility "in their backyard."

Other European Countries

Belgium, Finland, the Netherlands, Spain, and Switzerland have a combined total of twenty-six operating plants with a total capacity of 19.6 GWe. In a recent referendum in Switzerland, the nuclear option was confirmed by a 66 percent majority vote.[21] A number of European countries are implementing capacity uprates as a means of adding small amounts of nuclear capacity to their generating mix. For example, Spain has increased nuclear capacity at six units for a total of 600 MWe. Retirements of older, small-capacity units are to be offset by further uprates at newer units. Uprates have also taken place in Switzerland and Finland. The Belgian government has enacted phase-out legislation limiting unit operating lifetimes to forty years. The parliament in Finland recently approved the building of a fifth unit to be sited at one of two existing plant sites. First operations of a unit rated between 1,000 and 1,600 MWe is expected by 2010. The collective capacity of these five European countries is projected to decline to 18.3 GWe by 2020.

The Finnish company, Teollisuuden Voima Oy (TVO), has begun the site preparation work for the fifth nuclear plant, a 1,600 MWe advanced EPR. A consortium comprising Areva of France and Siemens of Germany will build the plant. Scheduled to be completed in 2009, it is expected to cost $3.7 billion in 2003 dollars. This reactor will help TVO to maintain its ability to supply nearly one-third of all the electricity used in Finland.

Korea

The Republic of Korea currently has nineteen nuclear power plants in commercial operation with a combined net capacity of 15.9 GWe, providing 39 percent of that country's electrical generation. Two additional units with a capacity of 1,900 MWe are under construction and scheduled to enter service by 2005, and eight additional units totaling 6.5 GWe are scheduled to be put into service by 2011. There could be 21.8 GWe of installed nuclear capacity at twenty-four units by 2015 and 24.8 GWe at twenty-six units by 2020. Growth in nuclear power is expected to take place with twelve new units coming online by 2020, as projected by the government. This expansion program would move Korea from sixth in world nuclear generation capacity to fourth by the year 2010.

There is some concern that planned capacity expansion is overly ambitious and may not fully reflect the 1998 economic crisis or about potential difficulties in raising capital once power market restructuring is completed.

Taiwan

Taiwan currently has six nuclear power plants in commercial operation with a combined net capacity of 4.9 GWe. Construction at the new two-unit Lungmen project continues. Previously, Taiwan's president attempted to cancel the units as promised during his campaign, and construction was temporarily halted. The nation's highest court subsequently ruled in January 2001 that the government's decision was unconstitutional because the president had acted without approval from the legislature, and construction was resumed. However, the schedule for initial commercial operations for the two plants was pushed back to 2006 and 2007, respectively. No impact has been observed from elections held in December 2001 that resulted in the antinuclear Democratic Progressive Party gaining a slight edge in seats over the pronuclear Nationalist Party, which previously controlled the legislature.

China

China currently has nine nuclear plants with a total capacity of 6.6 GWe, providing 1.4 percent of the country's total electric power generation. The completed units include Canadian, French, and Chinese designs.

Completed units include the two-unit 984 MW Dayawan Station and the two-unit 1,000 MW Ling Ao Station, both of French design, in Guangdong Province, and the 300 MW and the two-unit 600 MW Chinese design reactors in Qinshan Station, in Zhejiang Province. Under construction are two Candu 700 MW units, of Canadian design, in Qinshan, Zhejiang Province, and two units of 1,060 MW, of Russian design, in Tianwan, Jiangsu Province.

China began to experience electricity shortages in 2003. As well, the Sixteenth Communist Party National Congress stated plans to quadruple China's GDP by 2020. Due to the current constraints on electricity supply and the planned economic growth, China's National Development and Reform Commission, a national government-level organization, has revealed ambitious plans for a major expansion in its nuclear power program. The plan purportedly calls for approval of two nuclear power stations each year over the next decade or more.

China National Nuclear Corporation (CNNC) is currently considering a site in Sanmen, Zhejiang Province, for six units of nuclear reactors. China intends to standardize its future reactors to be PWRs and the size to be at least

1,000 MW. However, China is still undecided on the choice of technology, except that it would prefer Generation IV reactors.

Russia

Russia currently has thirty operating nuclear power plants with a total capacity of 20.8 GWe, providing 16 percent of the country's total electric power generation. The government's nuclear power growth scenario envisions a capacity of between 35 and 49 GWe by 2020. The government plans a substantial increase in nuclear capacity in absolute terms and a growing percentage of supply of electricity, mainly at the expense of natural gas (of which supply is expected to dwindle as the Siberian gas fields are depleted). The expansion of nuclear power in the short term is expected to be accomplished through lifetime extensions of its first generation reactors and increases in average availability factors from 75 to 85 percent.

There will be modest net growth for Russia over the next five years as the halted units are completed. The growth should level off as the retirement of first generation VVER-440 and RMBK units are offset by the completion of additional deferred units and the introduction of new designs such as the VVER-640. It is possible that Russian-installed nuclear capacity will increase 18 percent to 24.5 GWe in 2007 and range between 24.6 and 25.9 GWe thereafter.

Russia is taking measures to try to step to the forefront of world leadership in both the front and back ends of the commercial nuclear fuel cycle. Russia has been exporting enrichment services since 1973 and began exporting uranium in 1990. At the end of the last century, Russian annual exports of nuclear fuel cycle services exceeded $2 billion. It is pursuing a vigorous reactor sales effort abroad, with sales concluded in China, Iran, and India. There is also a proposal on the table for Russia to store and potentially reprocess foreign spent nuclear fuel, involving up to 10 percent of the world's existing spent nuclear fuel, up to 20,000 MT, to raise an estimated $20 billion in hard currency. The Russian government plans to put at least two-thirds of this income back into the expansion of its domestic nuclear capacity.

Ukraine

Ukraine has thirteen operating nuclear power plants with a net capacity of 11.3 GWe, providing 46 percent of its electric power production. The two plants which are under construction are greater than 80 percent complete and are projected to begin commercial operation during the next three years. To complete the Khmelnitskiy 2 and Rovno 4 PWRs (K2/R4 project), Ukraine is attempting to employ a combination of G-7 and Russian funding.

Announcements by the government in 2002 stated that there would be no more plans for developing nuclear power beyond completion of the K2/R4 project. Most forecasts assume that the K2/R4 units will be completed by 2006, raising total nuclear capacity to 13.1 GWe, where it may remain for the foreseeable future.

Ukraine's dependence on Russia for fossil fuel imports provides a strong incentive to keep nuclear generation high. Problems with Russian nuclear fuel delivery schedules, price, and quality have resulted in efforts to establish an independent fuel fabrication capability, although Russian technology will be employed to keep investment costs down. The desire to foster energy independence must be balanced against the realities of capital requirements for new nuclear power plants, which are projected to be beyond Ukraine's reach for the mid-term.

Canada

Canada currently has seventeen operating nuclear power plants with a combined net capacity of 12 GWe. Canada's nuclear generating capacity provides approximately 13 percent of its total electricity generation. Installed capacity is projected to increase to 13.6 GWe by the end of 2006, where it is currently expected to remain through 2020. In March 2004, the Manley report on Ontario's electricity industry recommended construction of more nuclear power plants to meet current and future demand.[22]

The Canadian nuclear industry in Ontario was hit badly by technical and management problems in 1997 and is only now beginning to emerge from them. The decision was made to lay up the seven oldest units at the Pickering and Bruce sites (one Bruce unit had already been taken offline in 1995) so that management could concentrate on improving performance at the remaining fourteen units. While there is still some room for improvement, as a result of these actions, capacity factors at the remaining plants had increased to 84 percent by 2002. Following an environmental review, Ontario Power Generation received approval to return the four Pickering A units to service in 2001, but schedule delays and cost increases continue to plague the restart efforts. British Energy has sold its share of the Bruce Power project to Cameco and the other partners. Bruce Power restarted Bruce 3 and 4, concluding a $550 million refurbishment program. While the provinces of New Brunswick and Quebec each have one operating plant, they have no plans for additional plants at this time. Atomic Energy of Canada Ltd. is currently pushing for the construction of eight new units over the next twenty years. Ontario's Independent Market Operator has warned that the province needs 11,600 MW of generation from new supply, improvements to current plants, and conservation in the next ten years.[23]

Other Countries of Note

Other countries of note with nuclear power programs as of 2003 are presented in table 4.4, which shows projected nuclear capacity for 2020. With the exception of India and the Czech Republic, most of these programs are relatively small, though that does not mean they are not significant sources of electricity. For example, in the Czech Republic, Deputy Minister Martin Pecina has announced a draft version of that country's energy plan to 2030, in which it is stated that the Czech Republic will increase its dependence on coal and nuclear plants for energy. "The most acceptable scenario for us is the one which sees up to 45 percent (of power consumption) to be satisfied by nuclear energy." This could be accomplished by adding 2,000 megawatts of capacity at the Temelin nuclear plant site.[24]

Table 4.4.   Status of Relatively Small Nuclear Programs

Country		Operating Nuclear Plants	
	Units	2003 GWe	2020 GWe
Argentina	2	0.9	0.6
Armenia	1	0.4	0.0
Brazil	2	1.9	3.2
Bulgaria	2	1.9	2.9
Czech Republic	6	3.5	3.5
India	14	2.7	20.0
Lithuania	2	2.8	0
Mexico	2	1.3	1.3
South Africa	2	1.8	1.8

Source: Based on World Nuclear Association fact sheets.

In India, the power secretary R. V. Shahi reaffirmed the government's policy to increase nuclear capacity to 10 GWe by 2012 and then add another 10 GWe by 2020.[25] India has nine plants either under construction or about to be. One of the plants that is scheduled for construction is a 500 MWe fast breeder reactor, a technology that India plans to widely deploy in the future because of uranium resource inadequacy and the limitations placed by other countries on its importation of uranium.

Regulatory Roadblocks
to Nuclear Power

One of the obstacles faced by the United States' nuclear power program over the past twenty years has been the regulatory system. Not only did the NRC stray from its basic mission, the judicial branch of the U.S. government handled technical matters that should have been referred to the NRC. There is no escaping the proposition that the United States, like no other country, created a system of multiple opportunities for court appeals on top of NRC licensing procedures. Action after action was taken, with the end result that no final decisions were truly made. Incredibly, it took up to fourteen years in some cases to build a nuclear power plant in the United States, when countries such as Japan and France were building them with the same high safety standards in six years. That was not the intention of the licensing process created by Congress.

Regulatory delays significantly contributed to increasing the capital cost of nuclear power plants as interest payments accrued over the interminable period of time it took the NRC to issue licenses. Overregulation led to staff increases that have cost consumers unnecessary expenditures. On top of that, for plants already licensed, the NRC forced shutdowns for technical violations that did not directly impact the actual safety of the public, further wasting consumers' money. In the 1980s and much of the 1990s, the NRC was guilty of all of these things.

At that time, the NRC was not a reliable regulatory agency. For example, for seven years (1991–1997), the NRC debated the Louisiana Energy Services' (LES) centrifuge-enrichment plant license application, without ever coming to a final decision. Other license actions were taking unpredictable amounts of times. Fear of the unpredictability of the NRC licensing process was effectively killing any interest in license renewals, power uprate applications, or any consideration of future new construction.

In addition to regulatory delays, I have been concerned for some years now about the fact that there is a poor understanding of the health effects of low-level radiation. Radiation standards are now determined with the LNT model, a model that is based on linear extrapolations from a small set of very high dose and dose rate exposures. Because of this approach, U.S. radiation standards are in question since they lack a verified scientific basis, according to the consensus of recognized scientists.

Historically, federal agencies, especially the Environmental Protection Agency (EPA) and the NRC, have sometimes disagreed over how restrictive U.S. radiation standards should be, and have even set differing standards (which include varying limits on radiation exposure to the public). Because of the high costs that result from the use of the LNT model and the imposition by the EPA of conservative radiation standards, the American taxpayer deserves to know if the fundamental theory behind the standards is accurate and if its use is warranted.

I also believe that using the LNT model as a basis for our radiation protection standards does another disservice to the American public. The notion that there is no reasonable threshold for radiation exposure other than zero, or as low as reasonably achievable, unnecessarily scares the American public. This fear of radiation surely has an impact on the public's acceptance of nuclear power. Yes, we must protect the American public from harmful radiation; however, our policy must be based on sound science.

The NRC's Day of Reckoning

In 1996, the NRC was embarrassed by reports that the Millstone nuclear plant in Connecticut had been operating outside its "design basis" by unloading its entire reactor core into a spent nuclear fuel storage pool without waiting an appropriate period of time for cooling. In response, the NRC shut down the reactor. However, this decision was not universally accepted as correct. In the words Dr. Richard Wilson of Harvard University, "the Nuclear Regulatory Commission strongly exceeded its authority and caused vast unnecessary expense in their shut down of the Millstone reactors in 1996."[1] I agree with Dr. Wilson and many other observers that the shutdown of the reactor was not justified.

To counteract criticism that the NRC allowed such operation to go on at Millstone, the chairman of the NRC, Shirley Ann Jackson, launched an NRC offensive to root out "design basis" flaws at other plants. Within a year, the NRC had created a list of problem plants and had handed out significant fines on individual utilities.

I heard from various people—from NRC employees to utility employees— that the NRC had become far too focused on creating more regulations and far less focused on developing the strong safety ethic that remains essential for our nuclear plants. During the time that there had been significant improvements in operational measures of nuclear power plants, the NRC had dramatically increased the number of citations for minor infractions—almost tripling them since 1995. It was painfully clear to me that the NRC needed to move beyond its focus on "design-basis" regulation to a focus on performance and risk-based regulations, as the Navy had done in its nuclear submarine program.

Since the NRC required reauthorization in fiscal 1999, Congress had a perfect opportunity for reassessment. In the Senate Energy and Water Committee's report on appropriations for that fiscal year, I announced my intentions to cut seven hundred jobs and consolidate departments at the NRC that would have saved $50 million. The Senate report outlined my specific concerns with the NRC, stating:

> The Nuclear Regulatory Commission has been subject to six major reviews since 1979. . . . The reviews contain common criticisms, among them: the NRC's approach to regulation is punitive rather than performance based, licensees are forced to expend considerable resources on regulations that are not related to safety, the NRC is unnecessarily prescriptive, licensees fear retribution for criticism, there are no specific criteria for important NRC actions such as placing a reactor on the watch list, and the NRC focus on paper compliance is not related to and can distract from safety activities.[2]

The Senate Energy and Water Appropriations Subcommittee's efforts were mirrored by those of our House counterparts who proposed even less funding than my committee for the NRC's fiscal 1999 budget. The House's appropriations report for the fiscal 1999 energy and water bill instructed the NRC to reduce its workforce, reduce the regulatory burdens on licensees, streamline its adjudicatory process, and issue detailed guidance on how it planned to review nuclear plant license renewal applications within a period of two years.

In my view, the problems at the NRC resulted from a change in regulatory culture at that agency as evidenced by the fact that the safety performance of U.S. nuclear plants had improved significantly while in the same period of time, the NRC dramatically increased its imposition of civil fines and level-four violations. The NRC launched a review of nuclear plant design baselines that required exhaustive reviews of design calculation, electrical separation, 10 CFR50.59 safety evaluations, accident analysis documentation, historical plant operating records, and steps taken to implement NRC "generic letters."[3] Tremendous costs were imposed on reactor operators even though significant deficiencies were found at only a few reactors. More importantly, the NRC's

new interpretation of what constitutes design base information created uncertainty as to what it expected of the reactor operators. The NRC became
increasingly willing to regulate management over and above the safety of the
plants. Its informal practices clearly circumvented the legal requirements for
imposing regulations. This practice was especially acute with regard to the
NRC's imposition of backfit requirements that in my judgment could not be
shown through a cost-benefit analysis to result in substantial increases in the
public's safety.

Frankly, the NRC had for too long existed without an appropriate amount
of congressional oversight. We allowed its influence to expand—especially in
light of the fact that no other agency has been a proponent of nuclear power.
I became very concerned that many NRC directives were not promoting safe
operations, and certainly were not using risk-informed criteria. While many
of the NRC requirements had questionable impact on safety, their impact on
the price of nuclear energy was far more obvious. These concerns injected uncertainty into the future climate for nuclear energy in the country.

The answer was to move to risk-informed, performance-based regulation.
The NRC needed to review existing regulations to reform those that were
outdated, paperwork-oriented, or that consumed resources for compliance but
had no positive impact on safety. The NRC also needed to review its interminable adjudicatory process imposed by the Atomic Safety and Licensing
Board. Its appeals process was sorely in need of streamlining. As well, it
needed a total about-face concerning the exorbitant and unpredictable time
that it took for the NRC to review applications, the broad discretion exer-
cised by judges to give standing to application opponents, and the effort the
agency undertook to resolve issues, no matter how trivial or unrelated to
safety. Part and parcel of this move to risk-based regulations was defining and
articulating the level of safety to be consistently applied to each nuclear
power plant.

Complaints about the NRC had become deafening, and discouraging.
Chairman Jackson asked me for a meeting and I decided I would take advantage of the opportunity to get some things straight. NRC staff fueled the ten-
sion by saying that the chairman had claimed she would likewise use the
meeting as an opportunity to "set the senator straight."

On the appointed afternoon, my appropriations subcommittee staffer, Alex
Flint, was briefing me on the latest outside evaluation of the NRC performance,
including a recommendation to cut NRC personnel by one-third, in light of the
NRC's inability to get its critical work done. Just as important, Democratic Senator Harry Reid of Nevada, the senior Democrat on the appropriations subcommittee I chaired, had been briefed on the evaluation and recommendations.
He was of the same view regarding the problem and solution.

I was prepared with irrefutable facts about the agency's performance and held this evaluation in my hand when the chairman arrived. I also had a copy of the budget recommendations in black and white for the chairman to see that I was not bluffing.

When Jackson arrived, I asked Flint to leave the room and I went to work. "You should know that I have never threatened an agency head with funding cutbacks in all the time I have been a senator," I explained. "But I have outside recommendations to cut your agency's budget by a third in light of so many failures in recent years. I know many of the NRC staff have adopted an adversarial attitude toward the entire industry, but it has gone too far. You can see in the subcommittee's proposed budget that we are serious about the cuts."

Chairman Jackson sat across the desk in the fading light of the afternoon. She seemed stunned when she saw the subcommittee's budget proposal. The chairman wondered aloud, "You can't be serious?" But as it became perfectly clear that I not only had the backup but the budget cuts in hand, she asked for details. I gave them.

Chairman Jackson pleaded for time, asking just how long she would have. I gave it to her straight, saying that the budget cycle would allow me a few months at most to see some real change or I would propose the full funding cuts. I offered to watch with interest the agency's performance and return some of the cuts if I was satisfied. Chairman Jackson got up, left, and didn't look back.

This "tough love" approach was necessary, as such views of the regulatory problems plaguing our nuclear industry were widespread. Within the next year, in 1999, the Center for Strategic and International Studies (CSIS) released its report concerning the NRC's performance, which concluded:

> [A] more fundamental change must occur in operational procedures and the way in which the NRC discharges its responsibility. Risk-informed regulation must be accepted and utilized by all levels of the NRC organization. Clear, concise definitions of adequate protection must be developed and understood by the industry, NRC, and the public. The how-safe-is-safe-enough argument should no longer distract or dominate the regulatory dialogue. . . . Industry and the NRC should strive to work more closely in a more informal and constructive atmosphere and conduct open dialogue with the public to arrive at regulatory procedures that allow the NRC to discharge its mandate effectively without imposing undue economic penalties on the industry.[4]

I believe that the single most important CSIS report conclusion concerned the original mission definition of the Atomic Energy Commission, and subsequently the NRC, to provide "adequate protection" for public health and safety from nuclear reactor operations—for which a quantitative definition had not been sufficiently developed. As modern risk-assessment technologies using probability theories have been developed over the years, there is a greater potential to move away from earlier approaches that focused attention on every single detail of the plant, rather than focusing on the smaller subset of components and issues that truly are critical to safe operations. I felt that this latter approach would lead to enhanced safety, perhaps at lower costs, and would greatly minimize controversy among stakeholders in the licensing process.

Focusing more on the most important determinants of safe operations allowed the NRC to move toward a more precise definition of "adequate protection." The approach suggested in the report to define quantitative thresholds for several specific areas of key importance enabled significant progress. Safety evaluations using probability theory helped greatly as the NRC moved toward "risk-informed regulations." This new philosophy was very important in providing the safety regime desired by the public while also enabling optimized cost performance.

Changes in NRC Regulatory Practices

After my Senate report recommendations and my meeting with Chairman Jackson in 1998, she proved to me that she understood the problems and pledged to fix them. Based on that pledge, I dropped the cuts and watched her and the NRC perform.

Events in the spring and summer of 1998 marked the turning point for the NRC. Outside groups, industry, and watchdog government agencies had been pressing for fundamental reforms in the NRC for years. Many inside the NRC were advocating change. The meeting with Chairman Jackson brought things to a head. As a result, NRC streamlined its adjudicatory process, made improvements to its inspection process, and moved to risk-based regulations. Statements from Chairman Jackson reflect and highlight the changes made in NRC procedures. On two separate occasions she said:

> With the Senate and House Appropriations bills as a catalyst, our FY 2000 budget proposal will reflect an approach that accelerates many of our efforts leading to a revised regulatory framework. We believe that accelerating our efforts toward a risk-informed and, where appropriate, performance-based regulatory approach will both enhance our safety decisions and provide a coherent basis for our regulatory processes. . . . As I have outlined earlier, we are committed to examining broad aspects of our reactor inspection, enforcement, and performance assessment processes (as well as other programs), and we will make the adjustments needed to optimize our performance in those areas.[5]

I have made the theme of risk-informed regulation central to my tenure as the NRC Chairman. In fact, the Commission is committed to the goal of using risk information and risk analysis as part of a policy framework that applies to all phases of our nuclear regulatory oversight, including rulemaking, licensing, inspection, assessment, and enforcement.[6]

Since that meeting with Chairman Jackson, I've been very impressed with the NRC. They are now a solid, predictable regulatory agency. They are accomplishing license renewals on a reasonable schedule, and most important, they are sticking to schedules once they publish them. This doesn't mean that the NRC approves all applications—and that was never my intention; it just means that when the NRC says they will finish a job by a certain date, industry can depend on it. In an industry with big capital costs, delays can cost a fortune and regulatory uncertainty can exert a stranglehold on all progress.

The Aftermath of NRC Changes

Changes at the NRC helped immensely with the rebirth of interest in nuclear plants, leading to dramatically increased optimism about the future of the industry. The NRC changed from an agency that took forever studying an issue to one committed to focused action. These changes have been more profound than first imagined.

The regulatory environment changed in dramatic ways since the problem era and the NRC deserves high praise and great credit for these changes. It transformed itself from an "adversary" to the industry to an "appropriate regulator." NRC's implementation of an effective license renewal process with improved procedures brought about industry confidence in the fairness of the process. No other program has impacted more profoundly on the nuclear industry, both here and abroad.

As of December 2003, twenty-six U.S. nuclear power plants had received twenty-year license renewals and eighteen applications were under review; an additional twenty-four applications are expected in the next several years. Renewal applications are completed within thirty months. As of October 2003, three U.S. nuclear plant sites had applied for early site licenses.

Between 1997 and March 2003, the NRC approved ninety-two power plant uprates and it expects an additional thirty-five uprates by 2007. These reviews have been completed within twelve months. The approved power additions through March 2003 equate to 4,022 MW, for a total added capacity of 6,292 MW. The NRC has certified the following advanced technology reactors: General Electric's ABWR, Westinghouse's AP-600, and Combustion

Engineering's System 80+. The Westinghouse AP-1000 is currently in certification review while four others are in preapplication review.[7]

Radiation Standards Issues

Regulatory standards, not just the process, demand increased scrutiny and understanding. Based on a study conducted at my request by the General Accounting Office (GAO) in June 2000, U.S. radiation standards for public protection lack a conclusively verified scientific basis, according to a consensus of recognized scientists.

> Below certain exposure levels, the effects of radiation are unproven. At these levels, scientists and regulators assume radiation effects according to the "linear no-threshold hypothesis," or model, under which even the smallest radiation exposure carries a cancer risk. However, the model is controversial among scientists, and decades of research into radiation effects have not conclusively verified or disproved the model, including studies attempting to statistically correlate natural background radiation levels in the United States and around the world with local cancer rates. Research is continuing, including a promising 10-year DOE program begun in 1999 [I created this program at the DOE to address the issue], addressing the effects of low-level radiation within human cells.

The consensus view that the GAO encountered among scientists and in the scientific literature is that the research data on low-level radiation effects are inadequate to either establish a safety threshold or to exclude the possibility of no effects. Individual viewpoints differed. Some scientists and studies held that the data support the existence of a safety threshold—an exposure level below which there are no risks from radiation. Other scientists and studies held that there is no such threshold and there can be risks at even the lowest exposure levels. In addition, other scientists and studies noted that risks from low-level radiation are complicated and variable, depending on factors such as the type and amount of radiation involved, body organs exposed, sex of the person, and/or age at exposure. For example, some researchers hold that children and fetuses may be more at risk from low-level radiation than adults. Some scientists and studies held that there are considerable data to support the view that low levels of radiation can actually be beneficial to health—the highly controversial theory of hormesis. Proponents of hormesis argue that research indicating beneficial effects has not been adequately considered in the "consensus" scientific community.[8]

A great many scientists seriously question whether the LNT model, upon which U.S. radiation protection standards are based, is appropriate. Many suggest that data would support the hormesis model mentioned above, wherein benefits are derived from moderate doses of radiation, perhaps by stimulating cellular repair mechanisms within the human body. Many suggest that the constant exposure to natural background radiation has required the body to develop internal repair mechanisms.

In a July 22, 2003, communiqué the French Academy of Medicine signaled their lack of support for the LNT model. The academy made the following recommendation, which highlights their position:

> Encourage major research efforts on the topic of the mechanisms, and the eval-uation, of the health effects of low doses of toxins, both chemical and radioac-tive. *The Academy would like to stress that the estimates of the health effects of low radiation doses (below a new mSv) or of low concentrations of carcinogens by means of a linear-no-threshold relationship has no scientific basis.*[9]

The LNT model forces us to regulate radiation to levels approaching 1 per-cent of natural background radiation levels despite the fact that natural back-ground can vary by far more than a factor of three within the United States, for reasons related to altitude, building materials, geologic environment, and expo-sure due to plane flights, to name just a few examples. We now use standards that severely restrict exposure to low-dose radiation, even to the point that we expect all work to be done such that the absolute minimum possible dose is delivered with virtually no reference to costs involved. We spend over $5 billion annually to clean contaminated DOE sites to levels below 5 percent of background.

If these standards overestimate risks, they force us to divert funds from other, potentially more worthy national goals. Alternatively, if the standards underestimate risks, we need to invest still more in cleanup activities. Many companies' profits from these cleanup contracts are enhanced by the use of the LNT model, which unfortunately tends to build a constituency with a vested interest in maintaining the LNT model.

Basing major policy decisions on the LNT model has significant financial ramifications. In a speech presented at the Uranium Institute in London, Dr. Bernard Cohen, a well-known radiation expert from the University of Pitts-burgh, highlighted how much money is possibly being wasted because of our reliance on this faulty theory:

> It is estimated that in the USA, US$85 billion will be spent in cleaning up the Hanford site to avoid LLR [low-level radiation], and comparable sums will be spent on government operating sites at Savannah River, Rocky Flats, Fernald and several others. If the LNT is wrong and LLR is harmless, all of this money will be wasted. Some other areas where huge sums of money are devoted to avoiding LLR are:
>
> - Radioactive waste storage technology and repository siting.
> - Reactor accident safety. Even in the worst accidents, well over 90 percent of the calculated deaths are from LLR.

- Reduction in routine emissions of radioactivity from nuclear plants.
- Reduction of radon levels in homes.[10]

The LNT model is also used to infer that miniscule doses ("decades" below natural background levels) applied to large populations through mechanisms like transportation of radioactive materials accumulate to lead to some number of fatalities. Such inferences then lead to headlines trumpeting the terrible risks to which the public is exposed. This use of collective dose data to estimate health impacts on large populations has been discredited in the scientific community, but some groups continue to use it.[11] Rarely, if ever, are these risks placed in perspective against other risk sources. And the gigantic uncertainties in the LNT model and significant evidence contradicting the LNT model are almost never discussed. Thus, many of the antinuclear groups have vested interest in using the LNT model.

The role of many antinuclear groups has especially puzzled me. On the one hand many of these groups express great concern over emissions of pollutants from fossil fuel plants, both from the perspective of fouling the air and from concerns over global warming. On the other hand, the simple fact that must be obvious to them is that nuclear energy is the only source of completely clean energy that is available today to provide a wide-scale solution for these pollution issues.

Maybe the renewable energy sources that these groups favor will make the impact that they hope in decades to come, but the economics are not correct now, despite billions of dollars in federal incentives. If these groups would direct some of their effort into finding good solutions for nuclear waste disposal, addressing potential proliferation issues with nuclear technologies, and seriously reassessing and updating the LNT model, I would find it far easier to believe the sincerity of their stated goals. In short, if they would balance concerns about the risks of nuclear with serious discussions of its benefits, and then direct some effort to address the risks, the nation might be able to make some real progress in this area.

Unfortunately, in the 1990s the EPA only reinforced these fears by publishing documents that claim to calculate, to several significant figures, the radionuclide risk coefficients for specific organs from specific isotopes.[12] Given the uncertainties in the validity of the fundamental model, I don't understand how the EPA could claim to have enough detailed understanding of the effects of low doses of radiation to publish such a document.

Some scientists have asked that I play roles as extensive as convening congressional hearings to explore the basis of the LNT model or that I legislate radiation protection standards. I have not called for such hearings, despite my strong interest in this problem. A Senate hearing is not an appropriate place for the evaluation of complex scientific questions. Senators are not the ones with the special knowledge to make these judgments.

I have encouraged, on the other hand, creation of a research program, mentioned previously, within the DOE devoted to serious study of molecular and cellular responses to low-dose radiation. I am very hopeful that this program can couple experimental capabilities with information from ongoing programs, like the human genome project, to provide us with real understanding on which to base intelligent standards for radiation protection. Whether the answer is that the LNT model overestimates or underestimates the risks, the correct information is vitally needed so that cleanup and regulatory activities can be appropriately adjusted.

BEIR VII

No discussion of U.S. ionizing radiation issues would be complete without mentioning the BEIR VII review by the National Academy of Sciences (NAS).

The NAS Committee on Health Risks from Exposure to Low Levels of Ionizing Radiation—the "biological effects of ionizing radiation" (BEIR)— was formed in 1998 to analyze the large amount of published data since 1990 on the risks to humans of exposure to low-level ionizing radiation. The committee was given the task of considering relevant data derived from molecular, cellular, animal, and human epidemiological studies in its comprehensive reassessment of the health risks resulting from exposures to ionizing radiation.

Regulations governing how much ionizing radiation the public can be exposed to are set by a joint process involving the NRC and the EPA. A report on biological effects of ionizing radiation, known as BEIR V, was published in 1990. In the mid-1990s, BEIR VI dealt solely with indoor radon and not with exposures to uncontrolled releases of radionuclides from nuclear facilities. BEIR VII, the study begun in 1998, was hoped to be the basis for resolving long-standing discrepancies in regulations in this area between the NRC and EPA.

In the 1990s a significant body of new epidemiological research accumulated that needed to be reviewed by the NAS. The most significant trend during the period, according to the NRC, was to question the LNT dose-response relationship at low levels of exposure. The dose-response relationship is the scientific basis on which federal agencies have developed federal regulations over the past thirty years. The issue of a threshold or not was addressed in the EPA's draft federal guidance report titled "Health Risks from Low-Level Environmental Exposure to Radionuclides," published earlier in 1998. The report drew criticism from the NRC and myself. In a letter to the agency, I said the

EPA had relied on "speculative and controversial mathematical risk models" in determining the likelihood of cancer occurring as a result of exposure to very low levels of ionizing radiation. It is believed by many that my letter helped prompt the BEIR VII study.

The NAS study, when completed, will establish the basis for risk analysis and regulatory reviews of exposures to low levels of radiation for years to come. The results will undoubtedly be used retrospectively to review exposures to low levels of radiation from sources such as atmospheric testing of nuclear weapons, uncontrolled releases of radionuclides from nuclear facilities, and other sources of radiation where exposures occurred in the range of 1–100 millirems above natural background.[13] Sources include naturally occurring radon in rocks and soil, medical X-rays, and exposure to cosmic radiation. About 11 percent of our exposure comes from our own bodies from the decay of potassium and other substances that we eat. Variations in exposures to the background radiation occur with occupation, geographic location, lifestyle, etc.

Following schedule delays of several years caused by data inputs, the NAS planned to publish the BEIR VII report in late 2003. The committee expected to have analyzed the most recent epidemiology from the important exposed cohorts and to have factored in any changes resulting from the updated analysis of dosimetry for the Japanese atomic bomb survivors. To the extent practicable, the committee also was to have considered relevant data from the DOE's low-dose effects research program. The committee was then expected to propose a risk model for exposures to ionizing radiation based on their evaluation of these data. However, because of further delays in receiving data from Japan, the report is currently not scheduled to be released until the end of 2004.

The EPA's Office of Radiation and Indoor Air will then prepare a white paper revising its methodology for estimating cancer risks from exposure to ionizing radiation in light of this report and other relevant information. This will be the first EPA product to be developed as a result of the BEIR VII report, and EPA requests that the Science Advisory Board issue an opinion during the development of this methodology. The second product to be prepared will be a revised version of the documents "Estimating Radiogenic Cancer Risks" and its addendum, which will describe the updated methodology and present new age-, organ-, and gender-specific risk estimates.

Uranium Resource Issues

Uranium Supply Initiatives

When I was elected to the Senate in 1972, the uranium mining industry in New Mexico was robust. However, the upheaval in the domestic mining industry, and especially in my home state, that took place in the early 1980s indelibly marked my understanding of the ultimate importance of the adequacy of uranium resources for our nuclear industry. It's all about resources. Electricity does not come from the plug in your wall; it comes from fissioning uranium, and burning coal, oil, and natural gas. Without the resources, the engines grind to a halt.

The U.S. uranium mining industry was thriving during the 1970s and New Mexico was its leader. American companies such as Kerr-McGee, United Nuclear, Homestake, Sohio Reserve, Anaconda, and United Nuclear Corporation were big players in New Mexico. In 1976, just four years after I came to the Senate, New Mexico was producing 48 percent of the country's uranium while Wyoming produced 32 percent.[1] To give you an idea of just how significant uranium mining was to my state, in 1976 the mining industry provided jobs for 3,833 miners and the milling sector to 1,046 people.[2]

Things began to unravel in the 1980s. During my first year as a senator, New Mexico produced 5,464 tons of U_3O_8 (a very stable oxide form of uranium). Its production peaked at 8,539 in 1978, after which production began to slow. It rebounded slightly in 1980 to 7,751 tons but began to show serious signs of decline by 1981 when production dipped to 6,206 tons. In 1982, even before the delayed impact of the Three Mile Island accident could be felt, New Mexican production was down to 3,906 tons, less than half of its peak in 1978. By 1983, of the group of uranium companies operating earlier, the only uranium processing companies left in New Mexico were Kerr-McGee and Homestake.[3]

As reflected in the state of affairs in New Mexico, U.S. production of U_3O_8 concentrates hit its high-water mark in 1980. Then cutbacks in mine and mill production began countrywide.[4] This industry contraction was partly the result of increased availability of lower cost foreign uranium supply. (Does this scenario sound familiar?) Procurement of foreign uranium for U.S. civilian end use was embargoed until 1977 when 10 percent of foreign uranium could be imported for use. However, by 1984, the year the embargo ended, domestic utilities took delivery of foreign uranium for almost half of their requirements.[5] The U.S. industry's doldrums were further exacerbated by reduced nuclear demand outlook. While nuclear power demand forecasts began to decline in the second half of the 1970s, the Three Mile Island incident further caused a retrenchment in the outlook for nuclear power and resulted in costly upgrades to operating plants as well as time-consuming delays to plants coming online.

The United States had been the world's leading producer of uranium, and until 1985, New Mexico was the country's leading producer. So as things changed I became more and more concerned and involved in the miners' situation, particularly those in New Mexico.

I responded to their plight by pushing for uranium industry monitoring legislation that resulted in Public Law 97-415 being enacted on January 4, 1983. This law established requirements for the president to annually review and report to Congress on the current and projected status of the uranium-mining industry, for DOE to develop a process for assessing the industry's viability, and for DOE to provide an annual "viability assessment" for each year in the 1983–1992 period.

During the next several years I found myself increasingly involved in domestic uranium mining and associated nuclear fuel issues, drawing me more and more into the broader issues related to the nuclear power option. While the first viability assessment under the monitoring law found the domestic uranium industry to be viable in 1983, all of the subsequent annual assessments through 1992 found the industry to be nonviable. It became increasingly clear to me and my staff that the miners' problems were also due to the poor nuclear demand outlook for the utility industry and that both sides had issues that needed to be resolved. This led me to believe that comprehensive legislation might be necessary to resolve matters for all of the parties in the nuclear industry, not just the mining sector.

I proposed a bill during the first quarter of 1986 entitled "Uranium Revitalization and Tailings Reclamation Act of 1986." Introduced on April 8, 1986, and reported out of the Energy and Natural Resources Committee on September 11, 1986, it had four provisions that were intended to solve most of the problems in the front end of the fuel cycle. It addressed uranium revitalization through either voluntary or mandatory import constraints, the establishment of a mill-tailings reclamation fund for the cleanup of twenty-seven specified mine sites that had provided supply under former government programs, invalidated the DOE's enrichment contract that the miners felt was the source of much of their troubles, and called for the development within one year of alternative methods of managing the government's enrichment enterprise. While this bill died in the 99th Congress, we reintroduced it in the 100th Congress.

Employment for current miners (my first priority), however, did not diminish the need to help those miners who had worked to provide uranium that enabled America to have the "nuclear umbrella" so critical to our security during the Cold War. So, we acted on two fronts: helping the industry to the extent we could (as I just described), and helping miners who suffered ill health through past work in dangerous conditions.

On the miners' health issue, litigation in our nation's courts posed complications. It was not until fiscal 1992 that we were able to finally get funding through the Radiation Exposure Compensation Act (RECA) that I had originally sponsored more than two years earlier. Even then, I had to propose funding in a national defense bill in fiscal 1992 to get actual appropriations flowing to the program.

The entire RECA program has run in fits and starts, unfortunately. The purpose of the program is to pay benefits to uranium miners, federal employees, and federal contract employees who participated in aboveground nuclear tests and for "downwinders" from the Nevada test site, if such individuals had suffered ill health from their work. The groups covered were those who worked for the nation's atomic energy program between the mid-1940s until 1971, and survivors of those workers. While we appropriated money through the Defense Appropriations Bill route, the program is administered by the Department of Justice.

And, confirming the rule that if something can go wrong it will, the Justice Department in May of 2000 began to run out of money to pay claims under RECA. Indeed, the Justice Department started sending out notices to claimants, those already approved for benefits, that since no money was available, it would issue "IOUs" to the claimants! The Clinton administration was largely responsible for the problem when it expanded the program, and then refused to ask for enough money to pay for legitimate claims. It took almost a year of hard legislative work to get the program back on sound fiscal footing, accomplished with the signing, by President George Bush, of the fiscal 2001 Supplemental Appropriations Bill in the summer of 2001.

Ironically, as this book goes to press, the Bush Justice Department has announced that once again, it may run out of money to pay claimants. So, the fight for justice for thousands of those whose health was irrevocably damaged by their role in the Cold War continues. I will never, however, let up on trying to get honorable treatment for these "veterans."

I recall becoming involved in a further contentious and somewhat related issue: the so-called unrecovered government investment in the national enrichment enterprise. Congress held that the "determination of the level of unrecovered costs that must be returned to the Treasury by the enrichment program . . . shall be made by the Congress in future legislation." Some members of the Congress supported a 1987 finding by the GAO that the amount was in excess of $7.5 billion, whereas the utility industry claimed that the amount owed to the U.S. Treasury, if not zero, was closer to $364 million. The DOE took the position that the amount owed was $3.25 billion.

Dealing with these issues led us to become more deeply involved in the question of restructuring the enrichment enterprise. Energy Secretary John Herrington advised the Senate Appropriations Subcommittee on Energy and Water Development in March 1987 of the conclusion that "the enrichment enterprise can best meet its commercial obligations and challenges if it is restructured as a federally chartered corporation which can be sold to the public through stock offering at an appropriate time." This is indeed the path that led to the establishment of the United States Enrichment Corporation (USEC) in July 1993, and then its privatization as USEC Inc. in 1998.

I had two concerns regarding the eventual privatization. First, there was concern about furthering our national nonproliferation goals if we were to give the stewardship of the Russian weapons-derived material to a profit-motivated entity. Second, the Clinton administration had provided large inventories of uranium to USEC. Those inventories were, in my opinion, likely to be used by USEC for their benefit at the expense of the domestic mining industry. My fears proved to be correct.

As previously mentioned, in October 1987, Senators Wendell Ford and Bennett Johnson and I introduced a bill entitled "Uranium Revitalization, Tailings Reclamation and Enrichment Act." The bill would have imposed a tariff on the import and use of foreign uranium in excess of 37.5 percent between 1988 and 1994, and in excess of 50 percent between 1995 and 2000. This bill also referred to the establishment of a U.S. government enrichment supply corporation, USEC, and of course, funded the reclamation of the mill-tailings sites associated with past government programs. Because of its proposed uranium import constraints, this bill was contrary to the administration's philosophy of free trade in general, and in particular, to the then imminent U.S.–Canada Free Trade Agreement. Some members of Congress at the time regarded the USEC title as tantamount to a subsidy because it would have "forgiven" the unrecovered government investment.

The execution of the U.S.–Canada Free Trade Agreement on January 2, 1988, ended the issue of uranium imports insofar as the domestic uranium mining industry was concerned. However, it did not end our efforts to find a solution for the poor circumstances of the domestic miners, many of whom were from New Mexico. In mid-1988, an administration- and mining industry-led proposal resulted in the so-called Uranium Package, which was to have been attached to the free trade agreement implementing legislation. While this proposal retained the tailings cleanup and the DOE enrichment enterprise restructuring provisions, it also proposed that the DOE build up a $750 million stockpile of domestically produced uranium beginning in 1989. The fact that the source of the funds for the cleanup and the stockpile was to be a levy on the utility industry's fuel use assured its early demise. However, an idea that I had at the time did survive the Uranium Package's demise—the Uranium Export Promotion program.

As part of our continuing efforts to push the omnibus uranium bill, and to see the export promotion idea come to fruition, I sent my administrative assistant and top energy aide, Paul Gilman, to Japan to review its nuclear program and to evaluate its efforts in the nonproliferation area. I subsequently visited Japan in the summer of 1988, shortly after the Congress approved the U.S.–Japan Agreement for Cooperation. With this agreement now in place, I encouraged the Japanese utilities to consider the purchase of uranium from U.S. mining companies, and as a result, the Japanese utilities agreed to buy a substantial quantity of uranium from the U.S. producers in the 1990s. Unfortunately, the promotion only benefited one U.S. mining company because the contract did not insist that the uranium must come from U.S. production. Most if not all of the uranium came from the spot market. A number of the U.S. companies may have substituted some lower-cost import supply for domestic production as the market price continued its decline in the 1990s. Thus, the initiative did not help the fledgling mining operations in the United States as was intended.

By the end of the 1980s I found myself continually involved in trying to weave legislation to enable the suppliers and the utilities to reconcile their nuclear fuel supply differences. Our efforts in this regard culminated in October 1992 in the passage of the Energy Policy Act (EPACT). This legislation provided for the establishment of USEC as a government corporation in July 1993 and for its possible future privatization. Insofar as the utilities were concerned, the legislation capped the cost to them of decommissioning and decontaminating the uranium-enrichment gaseous diffusion plants in the future at what seemed to be an acceptable total amount and forgave the "government-determined" unrecovered investment in past enrichment operations. This legislation also provided for limits on the entry into the market of the government's uranium stockpile materials and the authority to negotiate the purchase of Russian highly enriched uranium (HEU) and uranium feed derivatives. It also provided for mill-tailings cleanup and the continuation of enrichment research and development by USEC of AVLIS (atomic vapor laser isotope separation), the government's laser isotrope separation enrichment technology.

I found myself in the 1990s continuing to be involved in seeking solutions to issues between the Highly Enriched Uranium Agreement parties: USEC and the Russians and the administration. We wanted to see the HEU Agreement succeed because of its nuclear nonproliferation importance. The agreement had to succeed in my estimation because it would eliminate a huge arsenal of deadly nuclear weapons and provide funds for the employment of Russia's scientists so that they would not be attracted to clandestine Middle Eastern countries, such as Iraq, that were then seeking to develop nuclear weapons. It was apparent by 1994 that the miners did not want to see a flood of Russian natural uranium in the market and that USEC saw no value in being the marketing agent for Russian uranium, wanting only the enrichment component (which could be profitable).

Throughout 1994, 1995, and much of 1996, we fought to protect the miners' tenuous position and to preserve the HEU Agreement. On October 27, 1995, I wrote to Vice President Gore to ask him to reject an administration proposal for a waiver of the trade laws affecting the import of former Russian weapons material, believing "such a waiver to be unnecessary and unwise." I told him that the forward sales mechanism, established in the then proposed USEC privatization legislation, provided a market solution to the outstanding issues regarding the implementation of the U.S.–Russia HEU Agreement. In order to be successful, forward sales would require a predictable uranium market. It was our proposal and strong contention that forward sales conducted in accordance with the legislated schedule would not threaten injury, would be consistent with the national interest advanced in both the Department of Commerce Russian suspension agreement and the HEU Agreement, and would form the third leg of a stable uranium market.

Our hard work came to fruition with the passage of the USEC Privatization Act. Signed into law on April 26, 1996, it provided for the privatization of USEC, the disposition of U.S. and Russian HEU, and payment for HEU already delivered to the United States, and established an annually increasing quota for delivery into the U.S. market of the Russian HEU-derived uranium feed component. The act established that the Russians would be responsible for marketing the HEU feed component, though it did not address how. While we assumed that we had taken care of the HEU uranium feed issue and USEC business generally, we soon found out that there was more trouble

ahead in making sure the USEC privatization process did not undermine the
HEU agreement (covered in-depth in chapter 8).

The Uranium Supply Quandary

As we begin the new century, uranium resource issues once again move to the
forefront of the energy debate. *The question before us today is whether there are
adequate economic uranium resources to support expanding nuclear power programs around the world.* Demand for nuclear power has not just stabilized; it is
increasingly evident that demand for it will increase as the need for carbon-
emissions-free power grows, the price of alternative resources such as oil and
gas increases, and the availability of oil and gas peaks and then diminishes.
At current prices, nuclear fuel is only 26 percent of the cost of electricity,
whereas natural gas is 85 percent and increasing, it seems, on a daily basis,
making nuclear power increasingly economic. Nuclear power clearly will play
a larger role in the energy mix in the future. As this scenario comes into being, an increasing demand for uranium resources in the mid- to long term may
reframe the current fuel cycle debate. Consequently, there needs to be a discussion over the direction the United States should take with regard to the
current operation of light water reactors as well as the direction of advanced
fuel-cycle technology research and development.

In the remainder of this chapter, I examine the state of affairs of the
uranium-resource issue from the perspective of its supply, demand for the current generation of reactors and for the high-growth scenario, and conclusions
concerning the long-term adequacy of these supplies as compared to demand.

Uranium Supply Outlook

Uranium supply for today's and tomorrow's reactors comes from "already
mined uranium" (AMU) as well as from existing and future mines. The first
snapshot of the supply picture is of the estimated resources in the world, likely
sources for future mine supply, and the issues that impinge on industry's ability to mine the resources. Next the sources of AMU are examined.

The only comprehensive source of information is published jointly by the
International Atomic Energy Agency (IAEA) and Nuclear Energy Agency
(NEA); their estimates are published in what is known as "the Redbook."[6]
Table 6.1 summarizes the 2001 compilation of the NEA-IAEA resource data
categorized according to forward cost of extraction and certainty of existence.
Only the reasonably assured resources (RAR) and the estimated additional

Table 6.1. NEA-IAEA Redbook Uranium Resources

Category	Billion Pounds U_3O_8		
	<$15	<$30	<$50
Known:			
RAR	4.0	5.8	7.4
EAR I	1.4	2.2	2.8
Subtotal	5.4	8.0	10.2
Undiscovered:			
EAR II	0.0	3.8	5.8
Speculative A	0.0	0.0	11.5
Speculative B	0.0	0.0	14.3
Totals	5.4	11.9	41.9

Source: NEA-IAEA, Uranium 2001: Resources, Production and Demand (a joint report by the OECD Nuclear Energy Agency and the International Atomic Energy Agency), 2002.

resources (EAR I) are certain—that is, "known" to exist below the earth's surface. The data for the undiscovered resources, estimated additional resources (EAR II), are based on prognostication and indirect evidence and are therefore not very meaningful. For example, the United States reports all of its EAR as EAR II because the DOE has not done any geological resource data analysis since it closed its Grand Junction office in Colorado over twenty years ago. I understand that the fourth category of resources data in the Redbook, speculative resources, are even less meaningful since their location is unknown (!) and they are only surmised to possibly exist based on indirect evidence and geological extrapolation.

As I mentioned above, uranium resources are also categorized according to their estimated forward cost of extraction. As their definition implies, forward costs do not account for sunk costs, such as, for example, past unamortized costs. The Redbook cost categories are $15, $30, and $50 per pound U_3O_8. The $15 cost category applies to uranium that is found in today's spot market where prices of almost $20 prevail.

It can be seen in table 6.1 that only 24 percent of the Redbook-reported resources (10.2 out of 41.9 billions pounds U_3O_8) are known to exist. However, not all of these resources may be producible because of geological, political, and economic factors. There are other realities that impact resource availability. For example, while the 3.1 billion pound resources of the Olympic Dam mine in Australia are tremendous, it would take about 130 years to mine them at a rate of 24 million pounds per year, and thus, it would be well past 2050 and into the next century before all of this uranium becomes available. Since uranium from this mine is recovered as a by-product of copper mining, the rate of uranium exploitation will depend on the vagaries of the copper market.

Figure 6.1 shows where the RAR are located throughout the world for both the $15 and $30 reserve categories. Remember that the price of uranium at the beginning of this decade is in the lower cost category, throwing into doubt once again the issue of adequate economic uranium resources.

Four countries will provide about 90 percent of the Western world's uranium in this decade: Canada, Australia, Namibia, and Niger. Kazakhstan, Russia, and Uzbekistan will provide the remainder of the needed supply. The United States' northern neighbor, Canada, is the world's largest mine producer and will remain the dominant producer over the next several decades. At the beginning of the century it produced approximately 30 million pounds of uranium per year and may likely produce more than 40 million pounds annually by the end of the current decade. Australia comes in as the second-largest producer in the range of 20 million pounds per year. If expansions in license-approved capacity come to fruition as expected, Australia could increase its production to more than the 30 million-pound level.

Production from smaller suppliers may be afflicted with economic and political difficulties in some cases. Economic and currency issues could keep South African production at the current low level of about 2 million pounds per year for the foreseeable future. Annual production in Niger is expected to remain relatively constant during this decade at about 8 million pounds. There is some concern as to the continuing production of about 6 million

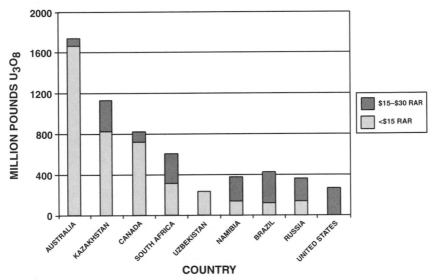

Figure 6.1. Distribution of World RAR. *Source:* Based on data from tables 2 and 3 of *Uranium 2001: Resources, Production and Demand* (a joint report by the OECD Nuclear Energy Agency and the International Atomic Energy Agency), 2002.

pounds per year from Namibia because of the low grades and associated economics of its single mine, and most importantly, currency issues relative to its neighbor, South Africa, to which it is financially tied. Poor market conditions are expected to result in China limiting production to meeting the needs of its own gradually increasing national nuclear power program. Production in Russia is projected to remain relatively constant at about 7.5 million pounds per year, even though Minatom had announced expectations for increased production. Kazakhstan currently produces roughly the same amount of uranium each year as its neighbor, Russia, but will likely increase its production by the end of the decade to at least 10 million pounds per year. Uzbekistan is expected to keep its production level at approximately 5 to 6 million pounds per year. Ukrainian production will remain steady at 2 million pounds per year and then will possibly face a phase-out due to economic constraints.

The most likely sources of uranium supply for the long term will be Australia, Kazakhstan, Canada, and Southern Africa. If the past is any guide to the future, Australian supply potential must be tempered with the reality that the official policy platform of the Australian Labor Party opposes nuclear power development. While current opposition could change as the environmental benefit of nuclear power is more widely accepted, this remains to be seen. Considering its size and future uranium needs, Russia must be considered as a prospective source of expanded uranium supply as soon as its embryonic uranium research institutions become better established. The United States may be considered as an expanding supply source for the long term because continuing and increasing demand will require that higher-cost category resources be developed in the future.

The United States' resource situation is unknown since a nationwide resource evaluation has not been undertaken by our government since the 1970s. The time has come for an assessment of our country's and the world's uranium resources.

With regard to the future and the potential for new mines to supplement the already known uranium mines, one must take into consideration the lengthy and costly environmental impact assessments that will be required before a uranium mine can be brought into development and eventual production. Because of the increasing complexity of mine development of many deposits around the world, particularly where high-grade deposits are involved, the deposits need to be large enough to support the overall economics of the projects. Because of the risks that are associated with mineral mining, particularly where radiation-emitting uranium mining is involved, the investors must be reassured as to the risks involved, and even then, they will inevitably demand high rates of return.

Another pressing concern regarding the assurance of adequate uranium supplies is the difficulties that will be encountered in identifying and bringing new supply sources into production. I am most worried about the current low level of uranium exploration and development activity around the world and the historically long lead times between reconnaissance, discovery, and production. The magnitude of the effort that will be required to provide a continuing supply of world-class production centers for the future is mind-boggling. New discoveries will likely be harder to find and probably deeper, even in regions that currently host economic production. In this regard, it will be increasingly in the consuming industry's interest to encourage exploration and development in regions that will assure them maximum diversity of supply.

As noted already, uranium that has already been mined will augment the world's supply. AMU falls into three broad categories: (1) commercial inventories; (2) HEU, the bulk of which is owned by the Russian and U.S. governments; and (3) uranium-bearing wastes, particularly enrichment tails and nuclear material-processing waste streams.

Tables 6.2 and 6.3 give estimated quantities of AMU in the United States and elsewhere in the world. The totals are estimated to be the equivalent of 534 million pounds and 1,143 million pounds, respectively.

Taking a look at the AMU in the United States, of the estimated 534 million pounds, approximately 16 percent belongs to the private sector. Of DOE's 70 million pounds of commercial-grade uranium, 42 million pounds are "frozen" as part of a 58-million-pound stockpile until March 2009 under the terms of an agreement with Russia that was intended to minimize the impact that this material might have on Russia's ability to sell HEU feed. Russia also agreed to maintain a similar-sized frozen stockpile until 2009. While the DOE owns an estimated 575 MT of HEU—containing the equivalent of 339 million pounds of uranium—it is not known if or when part of it will ever be released into the commercial market.

Table 6.2.   Estimated U.S. AMU as of January 2004 (million pounds U_3O_8)

Commercial inventories	85
DOE Commercial Grade U (excess HEU, NU)	70
DOE "Government Programs" HEU (based on estimated 575 MT HEU)	339
DOE Tails (<0.40 w/o)	40
TOTAL	534

Source: Various DOE reports.

Table 6.3.   Non-U.S. Estimated AMU (million pounds U₃O₈)

Non-U.S. commercial inventories	324
Russian HEU (899 MT) (115 delivered to the United States since 1995)	513
Russian commercial U and LEU	160
Russian enrichment tails (0.30–0.20 w/o)	96
Other U, LEU, and HEU*	50
TOTAL	1,143

*Excludes material owned by the Peoples' Republic of China.
Source: Estimates based on David Albright et al., *Plutonium and Highly Enriched Uranium 1996 World Inventories, Capabilities and Policies* (Oxford: Stockholm International Peace Research Institute, Oxford University Press, 1997) and other publicly available information.

Table 6.3 identifies the estimated 1,143 million pounds of AMU held outside of the United States. Adding the 324 million pounds of non-U.S. Western world and 160 million pounds of Russian commercial and low enriched uranium (LEU), the non-U.S. *commercial* sector AMU totals 484 million pounds. However, the reality is that this important source of supply today will decline over the next twenty years to the point where the bulk of supply will come from uranium mine production.

Russia is expected to blend down HEU-derived LEU containing another 228 million pounds between now and 2013, only about half of which will remain in the United States. It is thought and hoped by many, including me, that Russia may extend the original 1993 HEU deal for 500 MT of HEU by an additional 200 to 300 MT of HEU. By about 2020 or shortly thereafter, however, excess Russian HEU may be largely consumed, as will currently identified U.S.-excess HEU. The commercial availability of U.S. HEU in excess of the 174 MT of HEU identified in the mid-1990s as excess by the DOE will be subject to, and probably limited by, national security policies in effect at that time.

Uranium Demand Outlook

In July 2003, MIT published a nuclear power forecast (discussed in detail in chapter 4) predicting that worldwide nuclear-generating capacity might grow to 1,000 GWe by the year 2050 from a capacity of 353 GWe at the start of the twenty-first century. The corresponding uranium-demand forecasts for this growth scenario and for the current fleet of nuclear reactors are examined below.

As a "rule of thumb," a large current-design nuclear power plant (a 1 GW LWR) will consume approximately 30 million pounds of U_3O_8 during its assumed sixty-year lifetime. (Later, as we discuss advanced fuel cycles, we will see that advanced reactors can be far less demanding in their uranium resource needs and can utilize today's "spent" nuclear fuel.) The precise amount will actually depend upon many detailed factors including reactor design and fuel-management strategy. Nonetheless, if the world's installed nuclear power capacity rises to 1,000 GWe by 2050, as projected in the MIT study, then consumption between now and then, plus forward fuel requirements for the following thirty years, would amount to a total uranium requirement of about 28 billion pounds U_3O_8. This formidable requirement is almost three times the world's currently known high-cost uranium resources of over 10 billion pounds U_3O_8 as reported in the latest NEA-IAEA Redbook.

As an indication of the potential enormity of the uranium resource issue, it is of more than passing interest that the 350 million pounds of reserves and resources of the Cigar Lake mine in Saskatchewan, one of the two richest mines in the world (Olympic Dam in Australia is the other), is only large enough to support approximately twelve nuclear power plants throughout their assumed sixty-year lifetimes. About forty Cigar Lake-sized mines would be needed to meet fuel requirements of 1,000 GWe from now to 2050, and another forty Cigar Lake-sized uranium mine deposits would be required to assure at least thirty years of forward fuel supply for the projected 2050 nuclear power plant fleet.

A consideration of the uranium-demand situation for existing and near-term committed reactors shows that, based on current market conditions, overall world uranium demand is projected to remain essentially flat during the next twenty years, at about 178 million pounds per year. Then the outlook is for a gradual decline as the reactors run to the end of their lifetimes. It can be seen that approximately 5 billion pounds of uranium would be needed between now and 2050 under the current world nuclear power scenario. This picture is shown in cumulative form in figure 6.2.

The figure further shows that if we examine cumulative requirements for the existing and committed reactors through the end of their projected lifetimes or 2050, whichever comes first, it is evident that mid-cost ($30) resources (assuming that they are all actually producible) along with existing inventories of uranium (in their many forms) might be sufficient to supply the world's existing and committed reactors through their operating lifetimes.

Figure 6.3 compares the forward cumulative projected requirements to support MIT's projection of 1,000 GWe in 2050 through the end of the century against the NEA-IAEA known and undiscovered resources. The projections assume that the annual rate of growth of installed nuclear capacity will

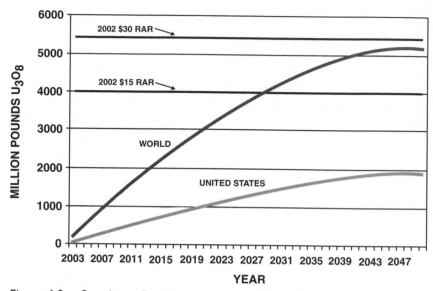

Figure 6.2. Cumulative Requirements for Existing and Committed Reactors to 2050. *Source:* Julian J. Steyn, "Uranium Resources: Need for Twenty-first Century Advanced Fuel Cycles." Nuclear Energy Institute's International Uranium Fuel Seminar, San Diego, California, October 12–15, 2003.

gradually decline from the 2.2 percent assumed during the first half of the century to 1.5 percent during the last ten years at the end of the century. Projections given have three requirements:

- actual requirements—operating requirements cumulated from 2003 to the year of interest,
- actual requirements to the year of interest plus fifteen years of forward requirements, and
- actual requirements to the year of interest plus thirty years of forward requirements.

The thirty-year forward requirements case assumes, for example, that at all times the industry would want to assure a forward resource availability in order to provide supply for the world's nuclear power plants fleet. The thirty-year forward resource inventory requirement in 2003 is about 5 billion pounds, and corresponds to the current $15 RAR plus EAR. By 2050, the thirty-year forward MIT scenario requires about 14 billion pounds to fuel reactors between now and 2050 and another 14 billion pounds to continue fuelling the 2050 capacity through 2080. Thus, in 2003, under the MIT scenario, we would need resources in the amount of 28 billion pounds to take us to 2050 and still have a thirty-year resource base remaining at that time, exceeding the known high-cost resources of ten billion pounds U_3O_8.

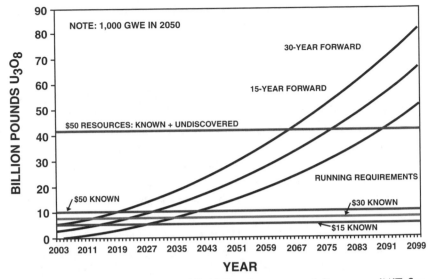

Figure 6.3.   Forward Cumulative World Requirements and Resources (MIT Scenario). *Source:* Julian J. Steyn, "Uranium Resources: Need for Twenty-first Century Advanced Fuel Cycles," Nuclear Energy Institute's International Uranium Fuel Seminar, San Diego, California, October 12–15, 2003.

While there should be adequate uranium resources to meet the needs of the current fleet of operating and committed reactors until well into the next decade, expanded nuclear growth as projected by MIT would require both an early expansion of the world's resource base and the early development of advanced fuel cycles.

Assuring Adequate Uranium Supplies for the Future

As a U.S. senator, I look at the uranium supply adequacy issue from a variety of perspectives. Not only am I concerned with the available supply, but also from whence the supply comes. I believe there is enough for the current generation of plants and that we can safely rely on the experts' projections of what will be required for the status quo. But as I believe that nuclear power is slated for a take off, huge uncertainties in the supply of uranium for the new capacity are added to the equation. I am concerned not only that there may not be enough affordable uranium resources to go around, but also that the United States does not currently have significant indigenous supply capability.

The current nuclear situation does not appear to cause concern regarding uranium resource availability. The cumulative requirements of existing and committed reactors worldwide will be approximately 5 billion pounds of uranium. Given conservative fact- and experience-based calculations, the economic resource base along with the AMU resources should be sufficient through 2050 to support today's nuclear power plants and those that may come online during the next several years.

However, we must seriously review the issue of uranium resource adequacy through the middle of this century from the standpoint of a nuclear power renaissance, and how it may drive the need and even accelerate the schedule for deployment of advanced fuel cycles. This subject is relevant today because existing and future nuclear power plants will require a supply of uranium fuel for their fifty- to sixty-year operating lifetimes. Cumulative uranium fuel requirements will continue to grow through the century and eventually will exceed even the currently known and speculative high-cost resources reported in the NEA-IAEA's Redbook.

While MIT's projected threefold growth in nuclear power is large, it is not unrealistic considering such a growth spurt would approximately maintain nuclear's share of world electricity generation at about 18 percent through the middle of the century. It would also serve to hold down the emission of carbon and other pollutants. However, if it is at all possible and because of the benefits it offers the world, the nuclear option should be expanded even further.

In order for the MIT scenario to become reality, and the environment to be protected from greenhouse gas emissions, there would need to be a tremendous effort to find and develop uranium resources. However, ultimately there will not be enough uranium fuel at reasonable cost to fuel MIT's growth scenario. As a consequence, in contrast to the assertions made in the MIT study, because of the long lead-times involved, there is an urgent need for the development and deployment of advanced nuclear fuel cycles in order to utilize our uranium more efficiently. The eleven members[7] that make up the Generation IV International Forum, including the United States, have already determined that there will be a need in the future for advanced nuclear energy systems and fuel cycles for several reasons, including, most importantly, uranium resource adequacy.

As we move forward in the development of nuclear power, the world's economic uranium resources will need to be doubled by 2020 and tripled by about 2035 in order to maintain a thirty-year forward resource inventory for MIT's projection. While the relatively low resource-finding rate of the past few decades was such because of the soft uranium market, it is going to have to more than quadruple to maintain an assured forward resource base. As a result, exploration and development must be significantly ramped up now. The last time there was a period of intense exploration activities was in the 1970s, when market prices were high. There may be a need for some type of incentives on the part of industry and government to spur exploration and development by the private sector.

I return now to my perspective on uranium supply, and my concerns regarding the security of supply for U.S. reactors. I cannot help but look at the potential of the advanced fuel cycle to augment the supply of uranium, which looks increasingly fragile in the nuclear growth case. I see a clear need for research now on advanced technologies in order to optimize the world's resources. We have a national treasure of uranium just sitting in our spent nuclear fuel storage pools. Each used rod is comprised of 95.6 percent uranium. The United States has already demonstrated at the Argonne National Laboratory that it can separate out this uranium at a high enough quality that it can be reenriched and manufactured again into new fuel for our current generation of reactors. When advanced fuel cycle technology is ready—and it is moving along but not quickly enough—it also promises us the ability to eliminate the plutonium left in the spent nuclear fuel rods by burning it, maximizing the power potential of already used uranium, as well as protecting the environment and the world from the dangers of plutonium. As we eventually move toward fast reactors in advanced fuel cycles that sharply reduce waste, even the residual uranium-238 in spent nuclear fuel becomes a usable fuel supply. This is a win–win situation. But the realities of developing technologies require that we undertake serious, focused research and development today to be able to give policymakers the options they may need by the middle of the next decade.

Revitalizing the U.S. Nuclear Infrastructure and Workforce

The nuclear industry is poised for a dramatic rebirth, ready to make contributions to solve our energy crisis. I am troubled, however, that there may be a dramatic shortage of trained experts ready to implement this growth. Without a robust university training system in the United States, as well as a productive, cutting-edge national laboratory program with the best and brightest minds on board, I am gravely concerned whether our infrastructure can meet the demands of a resurgent nuclear industry. Let there be no mistake, the decline in our infrastructure and the aging of our nuclear workforce pose risks to the future of the U.S. nuclear program and its ability to regain worldwide leadership.

There are many signs of our weakened nuclear infrastructure. For example, our nuclear workforce is graying, and the numbers of future nuclear engineers are declining. The universities are closing nuclear engineering departments, shutting down research reactors, and devoting fewer resources to information databases. Our national labs, world leaders in developing nuclear technology, have seen budgets slashed and facilities closed.

Experts have been worrying about this problem for the past several years. In response to the DOE's call for advice on managing its civilian nuclear energy research programs, the Nuclear Energy Research Advisory Committee (NERAC) was established in 1998. After completing numerous background studies, NERAC determined that:

- There is an urgent sense that the Nation must rapidly restore an adequate investment in basic and applied research in nuclear energy if it is to sustain a viable U.S. capability in the 21st century;
- An important role for the Department in the nuclear energy area at the present time is to ensure that the education system and its facility infrastructure are in good shape;

- The capabilities of currently operating DOE facilities will not meet projected U.S. needs for nuclear materials production and testing or research and development; and
- Of particular need over the longer term are dependable sources of research isotopes and reactor facilities providing high-volume flux irradiation for nuclear fuel and materials testing.[1]

Professor Michael Corradini, one of the country's leading nuclear engineering professors at the University of Wisconsin, identified the importance of these issues in a background report for the NERAC study: "A substantive and lasting investment in our human resource as well as our infrastructure is needed to enhance and provide for the public good through technology advances that support our nation's security, that supply its power and that contribute to medical advances, thereby enhancing human health."[2]

Even though many experts diagnosed the problem several years ago, the problems remain. As Dr. James F. Stubbins, University of Illinois professor and head of the Department of Nuclear, Plasma, and Radiological Engineering said at a June 2003 House Committee on Science hearing: "This period [1990s] saw the continued decline of several nuclear engineering departments and academic programs, and the loss of several critical university-based teaching, research and training reactors. This decline is still underway despite the current upward enrollment trends and increased research support for nuclear engineering program."[3]

Reversing the trends in engineering programs and reactor closures is crucial. They represent a fundamental and key capability in supporting our national energy policy goals in health care, materials science, energy technology, environmental cleanup, and national security. As Professor David M. Slaughter so aptly stated during testimony before the same House Committee on Science hearing:

> During the decline over the past several decades of student enrollments in nuclear engineering and radiation science programs, many universities chose not to replace faculty who left, which has created a shortfall of qualified faculty at a time when student enrollments are back on the rise. In addition, infrastructure neglect has occurred during the past few decades due to a number of complex issues, which include restricted budgets, increased costs of operation, the necessary diversion of resources to meet increased regulatory demands, and faculty turnover that may have resulted in changes to program directions. All of these factors and others combined with recent rapid technological and economic changes in nuclear engineering and radiation science leave many colleges and universities ill-equipped to impart the basic skills, interdisciplinary courses, industrial training, and modern and relevant research needed to better serve the industrial and government sectors.[4]

Several years ago I began to be fearful of the state of affairs in our universities and labs. We looked at the indicators for federal funding for the labs, the health of university nuclear engineering programs, demographics of the current workforce, and the attitudes of the future nuclear workers. We found some alarming trends.

Crisis in Our Universities

Nuclear science and engineering programs in U.S. universities faced a state of serious decline and are only now beginning a slow recovery.[5] For example, the supply of nuclear science and engineering personnel with a Bachelor of Science (BS) degree is still close to its thirty-five-year low. There were only 160 degrees given in the year 2000, a 20 percent drop from the previous year. The BS program enrollments declined from 1,400 in 1993 to about 500 in 1998.[6] Undergraduate enrollment in 1980 stood at 1,800; it declined to its low mark of 480 in 1997–1998, rising back up to 1,060 by 2002. Dr. J. P. Freidberg, professor and department head of MIT's Nuclear Engineering Department, also found a dramatic drop in undergraduate enrollments, from 1,852 enrolled nuclear engineering undergraduate students in 1979 to 570 in by 1997.[7] Two existing BS degree programs at the University of Maryland and the University of Illinois are under severe pressure and may not survive. Enrollment in undergraduate and graduate-level nuclear engineering programs is only half of its level ten years ago. The number of four-year-degree nuclear engineering programs has declined 50 percent to approximately twenty-five programs nationwide. At its zenith in the early 1970s, there were almost fifty nuclear engineering programs in our country.[8] The number of new faculty hires has diminished by 10 percent since the 1990s, and more than two-thirds are over forty-five years of age.[9]

Enrollments have declined for a host of reasons. The serious decline in university nuclear engineering program enrollments began after the Clinton administration energy policy, enunciated in the 1993 State of the Union address, devalued the role of nuclear power in the U.S. energy mix.[10] Figure 7.1 tracks how, as public perceptions about the nuclear power industry declined, student interest in nuclear studies seriously declined. An industry reported widely by the media to be "dying" was hardly in the position to attract the best and brightest students. There was intense competition from other industries vying for our young talented students, especially from the "sexier" high-tech companies.

The decline in federal dollars for research at universities and the financial pressures on the universities to close research reactors and other facilities fed

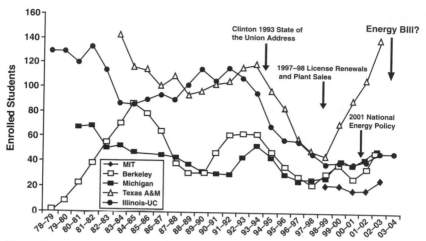

Figure 7.1. Student Talent and Interest Closely Track National Policy and Commercial Success. *Source:* Per F. Peterson, "Nuclear Energy in the United States in the Coming Decades" (speech delivered at the Atoms for Peace conference, Washington, D.C., December 8–9, 2003).

this precipitous drop in the number of programs and students. There were once as many as sixty-five university research reactors. By 1988 there were forty university research reactors; today there are only twenty-seven reactors. Lack of access to experimental facilities hampers student recruiting and leads to less well-prepared students. To compound this problem of access to research reactors, NERAC reported that within just a few years, 76 percent of nuclear scientists and engineers at the research reactors will be of retirement age, portending an alarming decrease in the number of teachers to train our students.

I lobbied Cornell University heavily in 2001 to keep its research reactor operating. But to no avail; it closed a year later. The Ford reactor at the University of Michigan closed in July 2003. MIT is regrettably considering closing its research reactor, which would be viewed as a real blow to the health of our university education program as MIT is considered the flagship nuclear engineering and research program in the world. Many more research reactor programs are endangered at our universities. On a more positive note, however, the research reactor program at Penn State University is in good shape primarily because the university spun off the reactor into a self-funding program and it does not drain the budget of the nuclear engineering program.

Our universities may not be able to continue to afford to maintain these reactors without federal assistance. It is estimated that a research reactor costs a university $25 million per year to operate. The DOE Office of Nuclear Energy annually provides very minimal funding for reactor refueling, instrumentation improvements, and reactor sharing for researchers. Most of the funding is spent on reactor refueling assistance. However, because many of the research reactors were built in the 1950s or 1960s, they will require relicensing and the inevitable new requirements, with subsequent increased needs for funding.

University reactor programs face yet another difficulty now that the NRC has imposed new spent fuel shipment standards affecting the universities. In October 2002, the NRC issued Interim Compensatory Measures (ICM) and orders to licensees, imposing additional security requirements to supplement existing regulatory requirements related to security for the transport of spent fuel in quantities greater than 100 grams. Licensees had already implemented additional measures after September 11, 2001. The orders and ICMs are not publicly available, nor is the list of NRC licensees who have received the orders; thus, the details of the ICMs are not publicly available. It is clear nonetheless that the regulations pose cost burdens on the universities. In addition, even though DOE actually owns the research reactor fuel, since the universities are the licensees that ship the spent fuel, universities are reluctant, without increased financial assistance from the government, to assume the increased risks in shipping imposed by the ICMs. I was pleased to take part in the decision of the U.S. Congress to lessen the universities' burden and provide funding, which began in fiscal 2004, for transporting spent nuclear fuel from universities to DOE facilities.

There are a few positive recent developments in the formation of new programs and departments and increasing nuclear engineering enrollments, which I attribute to the leadership President Bush has shown on the nuclear issue as well as increased emphasis by Congress. A sharp increase has occurred in student enrollments in most nuclear engineering programs.[11]

The University of South Carolina and the U.S. Military Academy have created the first new undergraduate nuclear engineering programs in the country in the past twenty years. Graduate programs are being created at the University of South Carolina as well as the University of Nevada at Las Vegas. Beginning in 1999, undergraduate enrollments at Texas A&M started to shoot up dramatically after the university engaged in a significant faculty and resource expansion.

Another notable development is the DOE's newfound commitment, as demanded by Congress and the Bush administration, to devote significant resources for university research. Universities are assisting in the development of advanced fuel cycles as discussed in chapter 8. DOE is funding university research in the areas of innovative fuels and materials, advanced separations technologies, transmutation technologies, and computation and modeling capabilities. DOE's involvement with our universities is a step in the right direction to assure the next generation of trained nuclear engineers, health physicists, and scientists crucial for our future. (However, this progress may be threatened by the Bush administration's fiscal 2005 budget proposal which, amazingly, proposes to reverse this trend.)

However, the state of the health of remaining nuclear engineering and related programs is still fragile. As the age of the faculty of engineering departments increases and undergraduate enrollments are still too low by university standards for maintaining a separate department, universities are trying to merge nuclear engineering departments or disband them altogether.[12] It is still too early to declare victory on the university front.

Nuclear Workforce Challenges

At the same time that the education and training of our future workforce is in decline, the need for nuclear industry workers is dramatically increasing. At this moment in history, the prospects for nuclear power are brighter than they have ever been since the early sixties. Companies are considering scenarios under which they might order new nuclear power plants, and several countries are joining the United States in R&D for new nuclear technologies and ways in which nuclear power could support a hydrogen economy. However, there are several factors that could stress the situation, including a current workforce that is close to retiring, as well as demographic sea changes as seen in the slowing growth of the U.S. workforce in general.

The nuclear industry is well aware of its workforce needs and shortfalls. Under the auspices of the Nuclear Energy Institute (NEI), a Work Force Issues Task Force was formed and a survey of the entire industry revealed some startling staffing problems for the future. The survey showed that under the current nuclear power scenario, the industry will need 90,000 new entry-level workers between now and 2011 for in-plant operations, government personnel and contractors, front- and back-end fuel cycle engineering design, services and construction, and for universities. The demand for degreed health physicists and nuclear engineers may significantly outstrip the supply for the next decade. The NEI study predicts that demand for nuclear engineers will be about 150 percent of supply over the next ten years.[13] Findings of the follow-up "2003 Nuclear Industry Staffing Survey" showed that the nuclear industry will need to fill 26,000 positions within the next five years primarily because of personnel retirements. This represents 46 percent of all jobs in the nuclear power generation sector.[14]

According to J. P. Sakey, senior vice president of TMP Worldwide, industry and government agencies alike face significant demographic challenges. In the past fifteen years, the labor force in the United States grew at an annual rate of 1.6 percent. Over the next fifty years, it is expected to grow by only 0.6 percent. With a U.S. population of up to 300 million, you can see just how dramatic the shortage will be of people entering the nation's workforce. By the year 2006, two workers will exit the workforce for every one entering.[15] The nuclear industry's challenge of finding an adequate number of trained nuclear engineers will be exacerbated by the general shortage of workers and the appeal of other employment opportunities.

Coupled with the predicted shortage of degreed and entry-level workers is the fact that the current nuclear workforce is graying. According to Ross Ridenoure of Omaha Public Power, one estimate shows that 30 percent of workers in the industry today can retire within five years, and half of the workforce can retire within fifteen years.[16] The NEI's "2003 Nuclear Industry Staffing Survey" finds that retirements will account for more than 60 percent of projected personnel attrition by 2008, nearly two and a half times greater than the current rate of retirement attrition. Approximately 17 percent of nuclear industry workers are between fifty-three and fifty-seven years of age, while 26 percent are between forty-eight and fifty-two, dangerously close to the average retirement age of fifty-five.[17]

The Government Accounting Office (GAO) prepared workforce studies for two of our key nuclear-related agencies, the NRC and the DOE. The GAO concluded that the shortage of technical staff at the DOE will reach crisis proportions in the next ten years. Of equal concern is the regulatory agency's future. The GAO study concluded that 33 percent of the NRC's technical professionals will be eligible for retirement by the end of 2005. That's just around the corner. NRC Commissioner Greta Dicus summed up NRC's predicament: "With a tight labor market for nuclear engineers and a workforce with a large percentage of personnel eligible to retire, the NRC is faced with some significant workforce challenges. I suspect that these challenges are not unique, and in fact, are shared with some other nuclear-related government agencies and with industry."[18] Who will fill their shoes with a declining population of nuclear engineering graduates at our nation's universities? This raises a red flag—how will the agency maintain the expertise and valuable institutional knowledge to effectively assure the health and safety of the nuclear industry?

The Declining Infrastructure at National Laboratories

The weakening of our university education and training infrastructure is mirrored in the decline in funding for nuclear research facilities at the national laboratories operated by the DOE. The mission of the national labs is broad—to provide a world-class research capacity to ensure the United States can meet its energy security and national security goals and advance the nation's knowledge in all sciences. The national laboratories have unique research facilities that integrate applied and basic research capabilities in ways that cannot take place on university campuses or in private-sector facilities.

The decline in the federal government's support for nuclear energy programs, which were ultimately terminated in the Clinton administration, has substantially weakened the national labs. In April 1977, before President Carter announced his "no reprocessing" policy, the United States was the world's leader in nuclear R&D spending. In the late 1970s, the U.S. budget hovered between the $2 billion to $2.5 billion level of annual expenditure. By 1985 it was almost reduced in half, to approximately $1.25 billion. In the first Clinton administration R&D funding was on life support and then was halted by 1997. Today, U.S. government R&D budgets for nuclear fission are significantly below those of France and Japan, and only marginally higher than the United Kingdom's.[19]

Directors of six national laboratories in the spring of 2003 presented Secretary of Energy Spencer Abraham with their assessment of the impact that the hiatus of nuclear power plant orders coupled with the decreased funding for nuclear programs has had. The directors wrote that among the three negative infrastructure trends that could in fact be worsened over the next decade are a: "weakening of the U.S. facility infrastructure for conducting nuclear research and responding to unanticipated problems. Many facilities within the university and national laboratory systems have been shut down, and the information infrastructure for sustaining nuclear databases and computational code centers receive inadequate support."[20]

There was considerable bipartisan concern over the decimation of our nation's nuclear R&D program. On July 30, 1997, I joined Senators Larry Craig, Frank Murkowski, Dirk Kempthorne, Jon Kyl, and Richard Durbin in writing to Secretary of Energy Federico Peña to argue for the reestablishment of an effective research program to further our national security, economic competitiveness, and protection of the environment. We wrote:

> The coming fiscal year will mark a notable event in the history of your agency and its predecessors. For the first time since the establishment of the Atomic Energy Commission more than forty years ago, the United States Government has no program to further the development of nuclear energy for the production of energy. This change, in the view of many members of the Senate, and in the view of this nation's energy experts, the lack of a strong and relevant nuclear energy research and development program, represents a major gap in the Department's energy research and development agenda.[21]

Thus was born the Nuclear Energy Research Initiative program at the DOE that has begun to stem the tide of indifference to nuclear matters that has had such a negative effect on our infrastructure and workforce.[22]

Congressional vigilance has been instrumental in repairing the damage. The Nuclear Energy Electricity Assurance Act of 2002, which I introduced in the Senate, contained provisions and significant funding for universities to promote greater nuclear engineering education. Another piece of legislation I cosponsored in February 2001, with Senators Bingaman and Crapo, is the Department of Energy/University Nuclear Science and Engineering Act. It highlighted the dismal state of affairs in the educational sector and offered solutions. The legislation would have significantly increased funding for graduate and undergraduate fellowships, junior faculty research grants, a nuclear engineering and education research program, communication and outreach activities, refueling research reactors and funding instrumentation upgrades; it would have provided relicensing assistance, created a reactor research and training award program, and increased university-DOE laboratory interactions. The needs remain today. The bill addressed the desperate state of our aging critical educational infrastructure that produces our nuclear engineers. The act aimed to authorize $238.8 million over a five-year period for university research and recruitment programs.

An inadequate R&D funding picture has not dramatically changed even though the Bush administration has recognized the important role nuclear power must play in ensuring America's energy security. For the 2004 funding cycle, the Bush administration has requested $127 million; Congress appropriated $130 million. For the next fiscal year, the Bush administration has inexplicably cut back funding requests to $96 million.

Prescription for Revitalizing Our Nuclear Infrastructure

We must turn the tide of the declining trends in maintaining our infrastructure. The preservation of our energy and national security demands no less. We must capitalize on key DOE programs now underway and seek new solutions. We must:

- provide adequate and stable funding for our university nuclear engineering programs and facilities;
- revitalize our U.S. national laboratories with adequate financial resources and appropriate management tools;
- support nuclear energy R&D to regain our international leadership on developing new nuclear technologies, fuel cycles, and missions; and

- encourage the private sector to work together with the universities to strengthen our educational institutions and their relevance to the workforce needs of tomorrow.

Passage of the provisions I had suggested as part of the Energy Policy Act would be a significant step forward in improving our infrastructure and workforce. It would strengthen university research and training reactors and their associated infrastructure. Specifically, it called for:

- converting research and training reactors using high enrichment fuels to low enrichment fuels, upgrading operational instrumentation, and sharing reactors among universities and colleges;
- providing technical assistance, in collaboration with industry, in relicensing and upgrading research and training reactors;
- providing the funding for university-owned reactor improvements;
- developing fellowship and visiting scientist programs between national laboratories and universities; and
- providing offsetting operations and maintenance funds for university research reactors.[23]

Support Our Universities

We must shore up university nuclear engineering programs and support the continued safe operation of our research reactors. It is imperative that we infuse the brain trust at our universities so that the professors can teach the nuclear engineers of tomorrow and perform the advanced research that will be necessary to bring advanced fuel cycles into fruition. The health of these programs depends in large degree on federal funding, although the private sector can play a role through research contracts at the universities and mentoring programs for the students. The DOE has several programs underway to provide fuel for research reactors, fund upgrades of these reactors, give scholarships, fellowships, and research grants to university students, and support university partnerships with the national labs and the nuclear industry. The DOE has committed to funneling a percentage of all nuclear energy R&D funds into the universities.

Strong support must be given by the federal government for fellowships, grants, and scholarships for students enrolled in science and engineering programs in U.S. educational institutions. Research grant funding has been relatively flat for the past five years[24] and must be increased not only in nuclear science, but also in the related fields of health physics and radiation safety, to say nothing of the desperate need for better support of all physical sciences.

Hundreds of highly rated research proposals are rejected each year because of inadequate funding. Scholarships for undergraduate and graduate students are woefully underfunded. With just over $1 million requested in the current budget and a cost of $55,000 to support a single student, you can see the disconnect between efforts to fund educational opportunities for nuclear engineers and other specialists and reality.

I believe one key to improving the pipeline of the highest caliber of trained personnel into industry, government agencies, and the national laboratories is to provide steady and adequate funding for the university research reactors. It is critical to maintain and improve the nuclear research reactors. As these reactors go through the license extension process, we must help the universities fund reactor and security upgrades that may be required by the NRC.

While it is not essential that every nuclear engineering program have an operating reactor, access to a reactor through a consortium is absolutely necessary to train our nuclear workers of tomorrow. I accordingly support research reactor fuel assistance and reactor upgrade funding. The regional university reactor consortia, established by the DOE at the direction of my energy committee and colleagues, must be expanded. These consortia must be geographically distributed. As well, I support the development of regional education and training consortia. Priority for federal assistance to universities must be given to those schools that maintain their research reactors.

Support Our National Laboratories

The number one priority is to increase R&D funding for the national labs to rebuild their expertise lost during the past two decades of severely reduced federal investments in our nuclear technology and expertise base. We must provide for world-class research facilities that can also support international cooperative research activities. The United States must protect its leadership role in advanced nuclear systems design, nuclear fusion, nuclear medicine, and nuclear space applications. Funding increases in R&D can be bolstered by increasing the ties between our universities and national laboratories. This can be accomplished through fellowships for graduate students to work at laboratories or with faculty exchange programs at the laboratories. Cooperative research through programs sponsored by the DOE, such as the Nuclear Energy Research Initiative and the Advanced Fuel Cycle Initiative, have expanded the university–laboratory ties.

Consistent with the goals and priorities set out by President Bush in his energy policy, the United States is embarking on a significant journey into new territories for nuclear energy. America will investigate how nuclear power can help us establish a hydrogen economy. We are developing advanced fuel cycles that make nuclear plants more proliferation-resistant, efficient, and more environmentally safe with regard to waste disposal. All of these initiatives will require the assistance of scientists in the universities and laboratories. The construction of new facilities for testing and demonstration will be necessary. Thus, new talent will be drawn into the nuclear field challenged and excited about the opportunities to share the limelight resulting from reaching new scientific frontiers. Our national laboratories' nuclear infrastructure will be revitalized in the process.

Our national laboratories cannot be adequately revitalized, however, until changes are made in the management of the national laboratories. Years ago, the federal government set out to improve the management of our national research institutions. Instead, we imposed so many mandates, policies, and procedures on them that today, the greatest minds in our country spend more time pushing paper than pushing the frontiers of human knowledge. We have blunted exploration and invention with bureaucratic policies and procedures. Science has been stifled by bureaucracy. We must find a way to refocus our laboratories on the scientific pursuits for which they were created and are generously funded. It is time to reexamine the fundamental management structure and relationship between the federal government and the national laboratories. As just two examples during the writing of this book, uncertainties about the management of Los Alamos National Laboratory and the Idaho National Laboratory are having a negative impact on recruitment and staff retention.

Support Private-Sector Efforts

There is a great synergy between the health of the industry and the vibrancy of our educational and research institutions. These institutions will flourish as the U.S. nuclear industry succeeds in deploying new nuclear power plants. But without a superior educational and research system to begin with, the industry will not have the technology base or highly trained personnel necessary for its resurgence.

The industry must continue to pursue the steps it has initiated to ensure an adequate workforce for the future. I commend the industry for creating task forces to quantify future workforce needs. This feedback to Congress and the government is essential for prioritizing funds for training where they are most needed. Industry must identify opportunities for federal grants, target funds on research contracts in the universities, and continue and increase the funding for scholarships and fellowships for university students, especially taking advantage of the federal matching grants program. The industry should consider financial support for universities willing to restart nuclear programs. Most importantly, the industry must help communicate to the students and leaders of tomorrow that the nuclear field carries exciting opportunities and is not a dying industry, but a robust industry whose technology will help contribute to alleviating poverty, war, and strife in the world.

8

Dealing with Nuclear Proliferation Effectively

Controlling nuclear proliferation[1] of all kinds—weapons, the uncontrolled diversion of weapons materials and equipment, and weapons expertise—is a challenge of vital importance. While it has always been a high priority of our government, the events of September 11, 2001, increased our motivation to maintain nonproliferation programs that provide effective international control of the materials, technologies, and know-how for nuclear materials or weapons that may be sought by terrorists or hostile governments in the future.

We also must effectively address any fears that nuclear power could be too dangerous to pursue in the current climate of terrorist threats and the nuclear weapons' ambitions of countries such as Iran, Iraq, North Korea, and Syria. The nuclear fuel cycle *can* and *must* be adequately protected and safeguarded. Pursuing advanced nuclear fuel cycles (both through closed and open fuel cycle approaches) *can* provide solutions to many of the potential proliferation concerns people may have about nuclear power. We need the courage of our convictions as well as sustained and consistent leadership on the nuclear technology front in order to make the promise of nuclear power without proliferation come true.

My vision of nuclear power rests on the belief, echoed throughout this book, that it is now providing, and can in the future provide a safe, secure, clean, and affordable supply of energy. If it is done right—with rigorous safety and safeguards protections—it is completely compatible with our all-important nonproliferation goals.

In the Name of Nonproliferation, We Lost Our Way

Stimulated, in part, by political reactions to India's test of a nuclear weapon in May 1974, America's twenty-year-old nuclear energy policy dramatically shifted away from pursuit of a coherent and widely accepted nuclear nonproliferation approach, including the appropriate pursuit of a closed fuel cycle. We started our descent away from our respected leadership position. Up until that time, U.S. policy was built on the belief that, if well-safeguarded, plutonium and uranium recycling, if pursued in some responsible industrialized states, would become advantageous from an economic perspective, conserve fissile uranium resources, minimize the need for uranium enrichment services, optimize the management and disposal of radioactive waste, and could be a good way to use and consume plutonium. However, after the Indian explosion, the U.S. Congress began to voice discontent with international nuclear policy and the performance of the International Atomic Energy Agency (IAEA). Some congressional leaders sought to blame U.S. policy and the IAEA for the Indian test—even though the Indians acquired their plutonium from an unsafeguarded Canadian-supplied research reactor.

Until that time, the United States was seen to be a leader in the nonproliferation area and a stable nuclear partner that believed in the orderly growth of nuclear power so long as rigorous safeguards and safety measures were assured. However, we started to move away from an international policy of cooperation to one of denial. The Ford administration was prompted to reexamine U.S. policy. Just before leaving office, President Ford announced that the United States should no longer assume that reprocessing of spent nuclear fuel for the production of plutonium was necessary or inevitable. However, President Ford did not abandon the idea that a closed fuel cycle could have merit. Rather, he was concerned that the utilization of reprocessing and of separated plutonium should not get too far ahead of the necessary safeguards and protective measures.

In April 1977, President Carter halted all U.S. efforts to reprocess spent nuclear fuel and develop mixed-oxide (MOX) fuel for our civilian reactors on the grounds that the plutonium in spent nuclear fuel rods from the nuclear power plants could be separated out, diverted, and eventually transformed into bombs. President Carter also took an aggressive stand that other industrialized countries should abandon their efforts to develop a closed fuel cycle even though many were committed to this course and the United States previously had said this was a reasonable course to pursue if well-controlled. People in the Carter administration and the Congress also moved to unilaterally change the ground rules for U.S. civil nuclear cooperation with other countries. This led to a major clash between the United States and several countries that was extremely damaging and from which we still have not recovered. The Carter administration decision to pursue the open fuel cycle may stand as one of the worst and most damaging in the last thirty years of U.S. nuclear policy.

President Carter, in his effort to remold nuclear polices worldwide, sponsored the International Nuclear Fuel Cycle Evaluation (INFCE) from 1977 to 1979. Forty countries participated in working groups and in evaluations assessing the relative proliferation resistance of different fuel cycles. After the INFCE, many of our allies concluded that all of the major fuel cycles had risks and benefits and that no one fuel cycle policy could be judged to be superior. They categorically rejected the Carter approach that pursuit of the closed fuel cycle should be abandoned.

INFCE expertise notwithstanding, President Carter "deferred" U.S. reprocessing and recycling programs. He effectively killed the development of such programs in the United States by shutting down the Barnwell reprocessing plant and the Clinch River Breeder Reactor. He argued unsuccessfully that the United States should halt its reprocessing program as an example to other countries in hopes that they would follow suit. While the Reagan administration revoked the "deferment" of reprocessing and plutonium recycling in 1981, by then U.S. utility and nuclear industry interest in these options had essentially disappeared.

Until the late 1970s, the United States was the nuclear fuel cycle technology leader. The U.S. Atomic Energy Commission operated defense program reprocessing plants at the Savannah River site in South Carolina, the Hanford site in Washington, and the Idaho National Engineering Laboratory in Idaho Falls. Between 1966 and 1972, Nuclear Fuel Services operated a reprocessing plant at West Valley, New York. It was shut down for modifications and expansion in 1972. However, substantial changes to federal regulations, particularly in regard to seismic requirements, resulted in the plant not operating again. While other reprocessing plants were planned and the Barnwell Nuclear Fuel Plant in South Carolina was half-built by 1975, they were doomed by the catastrophic changes in U.S. nonproliferation policy.

To begin to understand how we reached the point where the United States relinquished its influence on the development of civilian nuclear power around the globe and nonproliferation objectives, we need to recognize that the premises underpinning Carter's nuclear fuel cycle policy decisions were wrong. It is also important to recognize that the Carter administration essentially did much to create a political polarization of attitudes in our culture on desirable fuel cycle goals that has persisted to some degree ever since, and that has undermined our effectiveness to a significant degree. Prior to that time, pursuit of our civil nuclear and nonproliferation policies had enjoyed a substantial degree of bipartisan political support. The Carter era did much to kill that spirit and it led to distrust in other industrialized countries about the directions and stability of U.S. policies.

Until the advent of the Carter period, the United States had taken the position that it would be reasonable for nations to pursue either open or closed fuel cycles for their civil nuclear programs provided rigorous international safeguards, security measures, and other nonproliferation conditions applied. The United States also felt that sensitive technologies like reprocessing had to be closely controlled and not widely disseminated. However, their use was judged to be appropriate and desirable in well-defined conditions.

Carter administration staff, and some extremists, however, made the abrupt decision that these activities—even if well safeguarded—were inappropriate and should be abandoned. There was a fear that even if well safeguarded, civil nuclear plants and civil reprocessing plants could become a major avenue for the proliferation of nuclear weapons.

This failure to address an incorrect premise has harmed our efforts to deal with spent nuclear fuel and the disposition of excess weapons material, as well as our ability to influence international reactor issues.[2]

History shows that no country did what President Carter so feared would happen. To the best of my knowledge, no safeguarded civilian electric generating plant or safeguarded reprocessing plant has been clandestinely used for the production of plutonium for subsequent use in a nuclear weapon.[3] While it is true that a few countries have used "civil" activities as a cover for weapons programs, countries pursuing weapons programs have generally found it more opportune to acquire dedicated and unsafeguarded military facilities for this purpose. Moreover, to put this in perspective, several nations have judged it more in their interest to acquire HEU (by illicitly acquiring enrichment technologies) than to go the plutonium route via reprocessing.

The Carter administration was woefully wrong in thinking that the "American example" of killing the closed fuel cycle would inspire other countries with solid proliferation credentials to alter their intentions to operate and/or build new indigenous reprocessing or other sensitive fuel cycle facilities. Several other major nuclear power program countries—the United Kingdom, France, Japan, and Russia—all went down the reprocessing and MOX fuel road to plutonium management, leaving the United States with no seat at the policy-making table in discussing how the closed fuel cycle could best be managed.

The Europeans and Japanese elected to proceed with their reprocessing and plutonium-use programs. British Nuclear Fuels started construction of the Thorp reprocessing plant at Sellafield in 1983 and brought it into operation in 1994.[4] In Japan, the Power Reactor and Nuclear Fuel Development Corporation began operating its pilot Tokai-Mura plant in 1975. A subsequent large reprocessing plant went into construction in Japan in 1993, and is expected to begin operation in the next year or so. Moreover, France has successfully operated three reprocessing plants that went into operation in 1958, 1966, and 1990.

The argument that the Carter administration employed—that burying plutonium makes it safer from a proliferation point of view—in reality has just created unresolved issues for future generations. In fact many now agree that buried plutonium will become (as the radioactivity of fission products in the spent nuclear fuel is reduced enough to be palatable for excavation) an underground potential "gold mine" for would-be terrorists and nations looking for weapons materials.

Our country's pursuit since 1976 of a once-through nuclear fuel cycle will ultimately prove shortsighted. At a minimum, we have lost significant research and development time on spent fuel treatment technologies that could lead to better waste management and disposal scenarios that also could be more proliferation-resistant than current closed systems. As discussed in greater detail in chapter 9, reprocessing and recycling have the potential to greatly reduce the volume and toxicity of wastes to be disposed. At all times, the toxicity of the spent fuel is greater than reprocessed high-level waste in a closed fuel cycle. Clearly, advanced fuel-treatment technologies promise nonproliferation benefits as well.

Our country has lost not only R&D time to assess and develop better options for the future, but also influence on other countries' programs to develop more proliferation-resistant technologies and policies. We can devise better technical and institutional strategies, but only if we are in the thick of advanced R&D and are perceived to have a constructive and prudent role for nuclear power. On the issue of the growing spread of enrichment technology, we would have been far better off working more actively with our allies to stop the spread of it, rather than trying to dictate to other advanced countries with solid nonproliferation credentials what type of fuel cycle policy they should follow.

It is time to take stock of what the United States has helped to create. American leadership helped to build the global nonproliferation regime. From the outset, the United States took the lead in developing the rules of the road regarding international nuclear commerce that have become the basis of other countries' nonproliferation policies. We did much to help set the standard for acceptable behavior. In 1954 the United States amended the Atomic Energy Act to require that countries must have a bilateral nuclear cooperation agreement with the United States before an export of special nuclear materials or reactors to them could take place. Led by the United States, the international community undertook the commitment to halt the spread of the possession of nuclear weapons beyond those that already possessed them, as well as the spread of nuclear materials and technologies into the hands of terrorists. The international nonproliferation regime now includes a complex web of international treaties, national laws, bilateral and multilateral agreements, and international organizations and national agencies to stop the spread of nuclear weapons and other nuclear devices.

The United States was a driving force behind the development of the key foundation in the international nonproliferation regime, the Nuclear Nonproliferation Treaty (NPT), which came into force in 1970. The signatories to this treaty agreed that:

- nonweapons states will not acquire nuclear weapons and will allow international inspection of their civilian fuel cycle programs to ensure materials are not diverted to weapons purposes;
- the five declared nuclear weapons states—China, France, Russia, the United Kingdom, and the United States—may not aid other nations in developing nuclear weapons;
- the nuclear weapons states commit to the eventual disarmament of all of their nuclear weapons; and
- they also commit to collaborate with nonweapons states in the development of civilian nuclear energy programs.

Even before that, the United States helped build and nurture the IAEA, an arm of the United Nations that oversees the signatories' compliance with NPT obligations. Through its safeguards policies and systems, the IAEA inspects and verifies that nonweapons states do not divert nuclear materials for weapons purposes. A major component of the regime is the international group of supplier nations that have established guidelines for their "members" governing the export of nuclear materials and related technologies. In addition, there are international guidelines established by the IAEA to apply to physical security arrangements for the storage, shipment, and usage of nuclear materials.

The Nuclear Nonproliferation Act of 1978 (P.L.95-242) established the requirement that any non-nuclear-weapon country wishing to acquire special nuclear materials and equipment from the United States must first accept safeguards on *all* of its nuclear facilities. With the exception of China, a nuclear weapons state, all nuclear suppliers have adopted these standards. In retrospect, while the safeguards concept was admirable, we see that the United States' unilateral imposition of new requirements on our nuclear trading partners was similar to the Carter administration's attempt to dictate fuel cycle policies to other trading partners.

American leadership has been vital in the continual improvement of the nonproliferation regime. However, often in the process, the attitudes taken by some U.S. administrations (starting with President Carter's and to a lesser degree repeated by the Clinton administration) suggesting that we are hostile to nuclear power or theologically opposed to closed fuel cycles have impaired our effectiveness and influence.

Today, we are pursuing a more cooperative and engaged nonproliferation approach. George W. Bush spearheaded the Proliferation Security Initiative in the spring of 2003 that bands countries together with the common purpose of interdicting shipments of materials for weapons of mass destruction on land, sea, or air. He proposed in February 2004 a further tightening of the nonproliferation regime by restricting sales of reprocessing and enrichment technologies, requiring all states to sign the Additional Protocol (which has been ratified by the U.S. Senate), and formation of a special compliance committee within the IAEA to monitor member states' compliance with the rules.[5] These efforts are commendable and should be complemented by my recommendations at the conclusion of this chapter.

Building a Stronger Nonproliferation Regime

I have been one of the Senate's foremost champions of efforts to reduce the threats from weapons of mass destruction and nuclear weapons proliferation from the military programs of the former Soviet Union (FSU). (See appendix C for a chronology of nonproliferation events during my Senate tenure.) Frankly, foremost on my list of accomplishments in this effort was my role in the development of "cooperative threat reduction" (CTR) programs with Russia. This began with my support for the passage of the Nunn–Lugar Act in 1991, next with the creation of the lab-to-lab and materials protections programs, and then with the expansion of those programs in 1996 with the passage of the Nunn–Lugar–Domenici Act.

This list would also include my involvement in the privatization of the U.S. enrichment enterprise and passage of related legislation helping to create the U.S. program to purchase five hundred metric tons of Russian HEU. In the words of Charles Yulish, the vice president of the United States Enrichment Corporation (USEC), there is an "exquisite symmetry" to the fact that Russian bomb material has been used to light up homes across the American heartland. Over 90 percent of U.S. utilities have used uranium blended down from Russian HEU. I have been involved in efforts to ensure that USEC's privatization did not derail the HEU Agreement, but instead, supported it.

I am also proud of my efforts to establish the plutonium disposition agreement between the United States and Russia. I was not satisfied with the progress in this area begun in 1991 under the auspices of Nunn–Lugar cooperative ventures. In 1998, I developed a proposed blueprint for more rapid plutonium-disposition progress by working to convince the Clinton administration to incorporate a MOX fuel disposition option into a protocol for plutonium disposition that was signed at the 1998 Moscow Summit conference.

Since the initiation of these nonproliferation programs, which remain the lynchpin of U.S.–Russian cooperative ventures, I have helped lead the Senate's involvement and support of these critical national security programs. As a result, the United States has effectively removed highly enriched materials from thousands of weapons for possible rebuild into weapons within Russia, or worse— diversion to rogue states or terrorists. I am honored to sit with Senators Nunn and Lugar on the board of the Nuclear Threat Initiative (NTI), a foundation created by a contribution from Ted Turner in January 2001 to reduce the threat of the spread of nuclear, biological, and chemical weapons around the world.

This nonproliferation story does not end at the door of Russia. Events of September 11, 2001, have opened our eyes ever more widely to the emerging proliferation threats from a few rogue nations as well as from terrorists. We must deal with these threats, and we have made great strides with the widening of many programs originally established to help only Russia with the passage of legislation I proposed (the "Nuclear Nonproliferation Act of 2002") in the fiscal 2003 Defense Authorization Act.

We still face immense challenges. My long-standing and serious concerns for vastly improved controls over, and reductions in, tactical nuclear weapons remain unresolved. (While this is a real concern for me, nuclear weapons policy is not further discussed in this book on the future of commercial nuclear power.) I also worry that the amounts of Russian fissile materials that are not under effective control are still far too large. Some Russian weapons materials are stored in potentially vulnerable facilities that have yet to receive upgraded security measures. While we have funded promising programs for the conversion of Russian weapons scientists and the old closed cities to nonweapons and commercial endeavors, access to the closed cities is so difficult that progress has been too little, too late. I totally supported the CTR and other Russian initiatives, but believe, today, that U.S. programs must also focus on locations outside the Russian Federation.

In the post-9/11 world, we surely see the necessity of expanding our nonproliferation horizons to deal with proliferation threats around the globe. In many countries throughout the world, radioactive sources are poorly controlled and present significant risks for use in so-called dirty bombs by terrorists or subnational groups. We must also be concerned with the achingly slow pace of converting research reactors from using weapons-useable HEU to the use of low enriched fuels that do not pose proliferation hazards. The inventories of HEU in research reactors in Uzbekistan, Ukraine, Belarus, and elsewhere continue to be of concern.

Notwithstanding the proliferation issues, we must continue to hold out the promise of nuclear power to other countries, as envisioned in the NPT, in a manner consistent with our nonproliferation objectives. There is ample evidence that a successful nonproliferation policy requires a careful balancing of constructive cooperation in the civil nuclear sector as well as of controls and safeguards. Nonproliferation policies largely based on U.S. efforts to restrict delivery of weapons-useable material to other countries have been demonstrated to be unsuccessful. Unfortunately, it was the denial policy of the Carter era that did so much to hinder our effectiveness. Nuclear energy, appropriately designed to avoid proliferation concerns and operate in absolute safety, can—and must—play a major role in energizing the rest of the world. That is why we must develop a new generation of safe, proliferation-resistant nuclear reactors for the developing world.

Controlling Highly Enriched Uranium

The HEU Agreement

To go back in history a bit, I began to be very concerned that the U.S. enrichment enterprise's 100 percent Western-world market share at the beginning of the 1970s had been winnowed down to about half that by the late 1980s. The demand for U.S. uranium-enrichment services was falling as the two European enrichers were increasing their capacity to provide more and more of the world's product. So my staff and I began looking at privatizing the U.S. enrichment enterprise with the notion that because it was a government enterprise, by definition it must not be operating as efficiently as it would if it were owned by the private sector. Required to publish its costs and other sensitive information, it unfortunately provided the European enrichers with a competitive advantage. Without a consensus in Congress to go forward with complete privatization at the outset, I helped fashion a plan that was embodied in the October 1992 Energy Policy Act (EPACT). EPACT allowed for the establishment of the United States Enrichment Corporation (USEC) on July 1, 1993, to manage the U.S. enrichment enterprise, and created a timetable for taking USEC entirely private.

Closely following on the heels of the enactment of EPACT, the United States and Russia signed the historic HEU purchase agreement (the HEU Agreement) in February 1993. Also referred to as the "Megatons to Megawatts" program, the United States agreed to purchase 500 MT of HEU from dismantled FSU nuclear weapons over a twenty-year period. The purchased material was to be blended down by Russia to low enriched uranium

(LEU) and delivered to the U.S. executive agent for the agreement, USEC, at the port of St. Petersburg. The uranium feed and enrichment components of this LEU could in turn be sold in accord with U.S. law after delivery to the executive agent.

The deal's importance cannot be overstated. The United States agreed to take out of circulation enough nuclear material to build 20,000 nuclear weapons and to infuse $12 billion in badly needed money into the Russian economy in the process. As of the end of 2003, the HEU Agreement has resulted in the blend-down of 202 MT of HEU from approximately 8,050 nuclear warheads, 40 percent of the agreement's total. Most of the blended-down HEU to date has been consumed in U.S. civilian nuclear power plants to generate electricity for the American consumers. The HEU Agreement became the centerpiece of the efforts between the United States and Russia to reduce nuclear weapons and pursue nonproliferation goals.

Threats to the HEU Agreement

My staff and I have continuously sought to delicately balance the interests of the U.S. uranium miners with the necessity of maintaining support for the HEU Agreement. My two goals were to support domestic uranium miners by helping them maintain long-term supply contracts, while assuring the Russians would have a market for their blended-down uranium from dismantled nuclear weapons.

USEC, with its primary motive being to generate a profit at the time of privatization, threatened important U.S. nonproliferation advances made possible by the HEU Agreement. It was a constant struggle to keep the HEU Agreement from falling apart throughout the 1990s. At issue by 1996 was the fact that USEC only desired to buy the enrichment component of the HEU and not the uranium component. This agreement was jeopardized because the Russians needed to receive revenue for the latter.

The Russians began receiving uranium marketing proposals from a number of Western companies in 1996 and entered into negotiations with a group of three companies in June 1997. After on-again off-again negotiations, a deal was concluded between this group and Minatom in June 1998, only a month before USEC's privatization. But matters were not yet settled, since USEC's privatization revealed the DOE had endowed it with a tremendous uranium inventory. This was inventory, which the DOE previously had not been permitted to sell, available to the market from USEC! In its June 29, 1998, S-1 filing with the Securities Exchange Commission (SEC) prior to privatization, USEC went so far as to announce it planned to sell this inventory on what could only be described as an aggressive schedule. Many were concerned this would bring the uranium market price down to catastrophically low levels. We were especially concerned that this would jeopardize the HEU Agreement.

Even though I was a principal author of the privatization legislation, I was astonished to find that the soon-to-be-private company had a vastly greater uranium stockpile than envisaged in the privatization law. Within days, Minister Adamov, with whom I had developed a dialogue following his interest in my October 1997 "Nuclear Paradigm" speech at Harvard, was on the telephone to tell me that the agreement was dying.

Having been alerted to this very disturbing development, I began to let the administration know of my serious concerns that this development would have on budding U.S. nonproliferation initiatives in Russia. On June 26, 1998, three days before USEC's S-1 filing, I dispatched a letter to Sandy Berger of the National Security Council:

> In recent days, I have become concerned that aspects of the pending sale of the United States Enrichment Corporation, which serves as U.S. Executive Agent for the Russian HEU Agreement, depending upon the way in which that sale is structured, may have serious impact on implementation of the HEU Agreement and therefore national security. I want to ensure that the National Security Council considers those issues prior to a decision by the Secretary of the Treasury to approve the sale of the Corporation. . . . The Corporation's inventory of natural uranium may be significantly more than I understood to exist when Congress wrote the USEC Privatization Act.

On July 20, following a meeting at the White House with senior presidential staff, who were nonresponsive to my concerns, I sent them another letter:

> It is clear you understand the seriousness of the issues with regard to whether the inventory of uranium proposed to be sold with the corporation could imperil the HEU Agreement. If your remedy to the problem is to appropriate funds to underwrite the HEU Agreement, I may, because I so strongly support the objectives of the agreement, be forced to support your efforts. However, you would be much wiser to solve the problem by not creating it and reduce the amount of uranium transferred to the corporation before it is sold.

Not heeding my warnings, the National Security Council allowed the USEC sale to proceed and my predictions unfortunately came to pass—the HEU Agreement faced a full-blown crisis. At the 1998 summit I attended with President Clinton, I met face-to-face with Minister Adamov, who relayed his dismay over the developments. At the same time as Presidents Yeltsin and Clinton were announcing their plutonium safeguards and disposition agreement that I had worked so hard to support, Yeltsin threatened at this summit to pull out of the HEU Agreement. He had become frustrated both with the United States over the USEC-uranium inventory issue and the lack of agreement over payment terms for the natural uranium component of the LEU derived from HEU. Russian law stipulated that the material must be either paid for within six months or returned. The Russians considered this material to be a national asset. President Yeltsin's threat came as a huge surprise to administration officials—it created quite a stir at the summit!

Upon our return to the States, on September 8, 1998, I wrote to President Clinton proposing a compromise in order to save the agreement from collapsing. I proposed that the HEU Agreement might still be saved if the U.S. government would agree to buy the 1997 and 1998 Russian HEU-derived uranium feed deliveries that USEC had refused to purchase. My letter to President Clinton offered this compromise:

- Link U.S. purchase of 1997/1998 uranium to successful conclusion of a commercial deal for deliveries in 1999 and subsequent years (Russia's preliminary commercial agreement runs through 2006), removing any chance that the United States would have to step in again.
- Define terms for the return of uranium to Russia (directly or via Canada) and the permissible uses of it there (in FSU reactors or for HEU blending, prior to U.S. consent for other uses) or for its reexport to the United States in LEU.
- Based on satisfaction of these conditions, commit the U.S. government not to sell purchased Russian uranium or remaining DOE inventories for ten years.

Since my proposal, which had a price tag in the amount of $325 million, was not very well received at first by the administration, I took it to the leadership of the House and Senate. My proposal also called for the United States to define terms for returning the uranium component of the LEU to Russia for use in the blending down of HEU or as fuel in Soviet-designed reactors in Russia and elsewhere. Great credit is due to Congressmen David Obey and Bob Livingston and Senators Robert Byrd and Ted Stevens for their support in helping to save the HEU Agreement.

Ultimately, the president signed the budget bill on October 26, 1998, including the $325 million allotment. I sent my appropriations committee staff aide, Alex Flint, to Moscow on November 16, to make it clear to the Russians that disbursement of the funding was contingent on their concluding a long-term marketing agreement with the three Western companies in the near future. I followed him to Moscow about one week later to meet Deputy Atomic Minister Lev Ryabev and emphasize this reality, and let him know that the U.S. Congress would not provide any further funding.

With the U.S. government purchase of the 1995–1998 uranium feed now resolved, it was necessary to ensure that the Russians would be able to sell their future feed under an arrangement that was acceptable to them. With the assistance of the administration, the Russians and the three Western companies concluded the long-sought uranium marketing agreement on March 24, 1999.

The HEU Agreement's Future

Again in 2000, I became involved in trying to impress upon the Clinton administration the importance of supporting the HEU Agreement. On June 8, 2000, I wrote to Minister Adamov, and sent a copy of the communication to Secretary of Energy Richardson, stating, "I hope to emphasize to the [Clinton] administration the need to preserve the integrity of this key nonproliferation tool and develop approaches that will put the HEU Agreement back on a secure footing."

Problems had arisen because in November 1999, USEC had threatened to resign as executive agent if it did not get a subsidy in the amount of $200 million to manage the agreement for the next two years. It was clear to me at the time that USEC would be back for more funding in subsequent years. The stability of the agreement with Russia was threatened because of USEC's attempt to improve its then poor future financial viability outlook. During April 2000 congressional hearings, the possibility of a government takeover or buyback of USEC was raised.

During 2000 and 2001, USEC was in the throes of negotiating with Techsnabexport (Tenex for short), the Russian executive agent, the contract terms for HEU-derived enrichment purchases starting in 2002. While a contract agreement was given preliminary approval by the Clinton administration in its last days, the incoming Bush administration studied the issue and did not give its approval to contract terms until February 2002. Official U.S. and Russian government approval was announced on June 19, 2002. While the latest agreement between Russia and USEC seems to provide some stability, Russia continues to sell its enrichment component below market prices.

Today the challenge is that the HEU Agreement expires in 2013 and people are already discussing the merits of continuing or expanding the HEU Agreement. The enrichment component of the Russian HEU is now 50 percent of total U.S. sales by USEC and meets about half of domestic enrichment requirements. The Senate Energy and Water Appropriations Subcommittee, of which I am chairman, proposed in the fiscal year 2004 Energy and Water Appropriations legislation to provide $30 million to accelerate the purchase of Russian HEU in amounts beyond the 1993 U.S.–Russian HEU Agreement. However, the House failed to concur in this action. The House and Senate conferees merely directed the Energy Department to "develop a rigorous risk-based priority setting process for allocating budget resources to the activity with the highest nonproliferation benefit."[6]

Sometimes overlooked is the fact that a substantial portion of our uranium and conversion services supply is also obtained from the Russian HEU. As far as domestic nuclear fuel supply is concerned, we are down to one enrichment plant operating at half of its nameplate capacity, one conversion plant, and two very small foreign-owned uranium mines that meet only about 4 percent of our needs. I believe we must undertake a serious review to decide how much Russian HEU we want. Furthermore, it is essential to make this decision in the context of maintaining domestic enrichment and conversion capabilities for energy security reasons.

Nonproliferation Initiatives in Russia

Russian Policy Overview

In the 1990s, Congress focused heavily on the proliferation problems stemming from the dismantlement of weapons from the FSU after its breakup at the end of 1991. Through the creation of programs for securing and burning Russian weapons-grade materials, we have made huge strides in addressing some of our most pressing proliferation concerns. As I have indicated earlier, I believe we should now broaden some of these efforts to address some risks being faced in other countries.

After the breakup of the Soviet Union, together with other like-minded people, I set forth the following ideas to help ameliorate the proliferation concerns coming out of Russia. My main tenets in dealing with the Russian proliferation issue have been to advocate measures to:

- Better secure Russian weapons-grade material
- Convert weapons-useable material to unclassified forms
- Provide for safeguarded storage of excess weapons-useable materials
- Reduce Russia's weapons remanufacturing capability
- Recommend that U.S. strategic policy be based largely on the concept of parity—that is, similar numbers and types of weapons held by both the United States and Russia and on the level of the threats we face
- Push for programs to utilize MOX fuel in reactors rather than pursue the immobilization and disposal of weapons material
- Initiate cooperative programs with Russia
- Argue for the appointment of a special advisor and coordinator to the president for all U.S. nonproliferation programs

Pete Domenici had a close relationship with his parents, Cherubino and Alda Domenici. His mother's consent helped launch Pete Domenici's public service career after friends dared him to run for office or stop complaining about the Albuquerque city government.

Pete Domenici's great-uncle Lorenzo started the Montezuma Mercantile Co. in Albuquerque. The store was later bought by Cherubino Domenici, who transformed it into a successful wholesale grocery company. Pete Domenici as a youth worked for his father.

Domenici test drove the prototype of a hydrogen fuel cell car in February 2003 shortly after President George W. Bush unveiled a major hydrogen fuel cell research initiative.

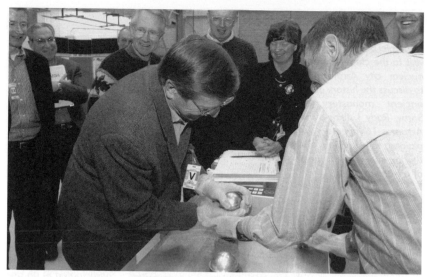

At Los Alamos National Laboratory, Domenici held a radio-isotope thermoelectric generator (RTG) designed to power space instrumentation through the heat generated through radioactive power. This is a small but important application of nuclear technology. Senator Harry Reid of Nevada and former NNSA administrator General John Gordon joined Domenici for the visit and RTG briefing.

An excellent demonstration of the French waste management strategy is this comparison between the total amount of waste generated by a French family of four over twenty years and a U.S. quarter coin. Reprocessing reduces the volume and toxicity of French wastes.

Domenici and former Senator Sam Nunn of Georgia continue to work together through the Nuclear Threat Initiative. In the Senate, the lawmakers collaborated on a variety of legislative initiatives.

In 2003, Domenici was honored to address the Woodrow Wilson Center Conference on the fiftieth anniversary of President Dwight Eisenhower's famous "Atoms for Peace" speech on December 8, 1953. (Courtesy C. David Hawxhurst WWICS)

Domenici works with his "nuke team" including chief of staff Steve Bell, science advisor and physicist Dr. Peter Lyons, and Alex Flint, staff director for the Senate Energy and Natural Resources Committee.

Pete Domenici has had a life-long love of baseball, playing ball throughout his youth. In 1954, he pitched for the Albuquerque Dukes, a farm club for the old Brooklyn Dodgers.

Cherubino Domenici emigrated from Lucca, Italy, to Albuquerque, N.M., in 1906 at the age of 14. He became an American citizen and fought in World War I. He later married another immigrant from Lucca, Alda Vichi.

A young Pete Domenici lines up with his three older sisters—Marianella, Mary, and Thelma—at their Albuquerque home.

The dismantlement of tens of thousands of nuclear weapons in Russia and the United States had left both countries with large inventories of perfectly machined classified components that could allow each country to rapidly rebuild nuclear arsenals. This led me and a few other observers to the conclusion that both countries should convert those excess inventories into non-weapons-useable shapes as quickly as possible. The more permanent those transformations and the more verification that could accompany the conversion of that material, the better. Since the early 1990s, I have repeatedly noted that the single most important step in lessening the proliferation threat from dismantled weapons is to promptly transform the material out of classified shapes and get it into internationally-monitored storage—and that primary objective remains relevant today.

Technical solutions exist to improve the situation. Pits can be transformed into nonweapon shapes and plutonium can be burned in reactors as MOX fuel—which, by the way, is one major option that the National Academy of Sciences recommended in the mid-1990s. (They suggested a dual-track approach that would have involved both burning plutonium as MOX fuel in reactors as well as immobilizing some of the material by mixing the plutonium with highly radioactive waste.) However, in the late 1990s, the proposal to dispose of weapons plutonium through burning it as MOX fuel ran into that old criticism (which started to surface with the Carter administration) that irradiating plutonium as MOX fuel is too risky despite its enthusiastic use by some of our allies. I, however, have argued and continue to argue that getting rid of our plutonium by burning it is the best technical solution for dealing with any excess inventories.

I was not opposed to the approach adopted by the Clinton administration that a dual-track disposition program (involving both the use of MOX fuel and immobilization) be pursued. At the same time, I did not object when, after an extensive review, the Bush administration decided to drop the immobilization route. This is an option that the Russians really never liked. Moreover, the Bush administration concluded that by dropping it, they could help reduce the costs of the program. So for now, immobilization is no longer in the picture.

I recognize that there are alternative approaches to plutonium disposition and I don't want to rule out the possibility that better programs may be developed to supplement our current path. As an additional approach, I've championed the study of new reactors for disposition of plutonium, with the hope that they could accelerate the rate of disposition of Russian materials. The high temperature gas-cooled reactor was jointly funded under this program between the United States and Russia for several years. While never brought to large-scale commercial use, that reactor has great potential to be an effective consumer of separated plutonium. Additionally, I've advocated careful study of thorium fuel assemblies as another option. I am encouraged that there is strong international interest in joint sponsorship of such new reactor approaches.

The Nunn–Lugar Breakthrough

While I've given the reader a snapshot of the thrust of my Russian policy, let's go back to where this opportunity began. After the failed August 1991 coup d'etat attempt against President Mikhail Gorbachev, Senator Sam Nunn, the chairman of the Senate Armed Services Committee, met with President Gorbachev. Realizing that the Soviet president had lost control over Soviet nuclear forces during the coup attempt, Senator Nunn returned to Washington to enlist the support of his colleagues to do something about the threat that Soviet nuclear weapons could end up in the hands of terrorists or other anti-Western governments.

I well remember the day when Senator Nunn brought up this issue on the Senate floor, on November 13, 1991. His proposal, the "Soviet Nuclear Threat Reduction Act of 1991," was quite controversial, and many were criticizing it as nothing more than another foreign aid program. I rushed to the Senate floor in support of my Democratic colleague with an impromptu speech, stating that we must realize America has a real interest in preventing chaos in the Soviet Union and must help get its nuclear weapons under control. In my remarks, I said:

> I came to the floor because I was listening to my good friend, the very distinguished Senator from Georgia, speak about the issue of what are we going to do, if anything, to help the Soviet Union. . . . Why is that important for Americans?
>
> I think things are dire in the Soviet Union. I do not believe they will make it through this winter without civil strife, without many, many, problems, probably without some civil disorder.
>
> Yet there are some who sit around and say, well, we will work with the Soviet Union later, just as soon as they get everything in place, their governance, a nice healthy democracy, and just as soon as they have capitalism going, we will work with them.
>
> I regret to say that this Senator believes that long before that date will come, there will be absolute chaos in that country, if not starvation, and who knows where all those nuclear weapons are going to be? And who is going to control them?
>
> I am very supportive of what the Senator from Georgia said here, and I compliment him for saying it and for acting on it. The easiest political thing to do is to be totally negative and to say the American people are not for it, so we are not going to do anything about reducing chaos in the Soviet Union. Frankly, I

do not think that is the case. I think our people would support some significant and orderly assistance during this transition. For some of us to stand up and say it is precisely what is needed, in my opinion.[7]

Believing in Senator Nunn's and Senator Lugar's idea, I worked tirelessly to help win passage for their bill. Ultimately they prevailed and the legislation created the Cooperative Threat Reduction (CTR) program. Senator Nunn has often mentioned to me that the true name of the act should have been the Nunn–Lugar–Domenici Act. While I appreciate his recognition of my efforts, regardless of its title, it was a great day for nonproliferation when the act was signed into law on December 12 by President George H. W. Bush. It carried an initial congressional appropriation of $400 million for fiscal year 1992 to secure and destroy the Soviet Union's weapons. Its three initial tasks were to:

- destroy nuclear, chemical, and other weapons;
- transport, store, disable, and safeguard weapons in conjunction with their disposal; and
- establish verifiable safeguards against the proliferation of those weapons.

The timing was brilliant—two weeks later, President Gorbachev resigned and the Soviet Union began dissolving.

Over the past decade we have tirelessly supported and funded programs in Russia undertaken by our Departments of Defense, State, and Energy. CTR programs administered by the Defense Department help Russia guard and control its nuclear weapons complex. The Energy Department now has responsibility for additional key programs such as R&D for the monitoring and detection of weapons activities; the Materials Protection, Control, and Accounting (MPC&A) program to track and control fissile materials; and programs to prevent "brain drain," such as the Initiatives for Proliferation Prevention (IPP), also known as the lab-to-lab program, and the Nuclear Cities Initiative (NCI). IPP is a cooperative arrangement between DOE labs and science and engineering institutes in Russia, Ukraine, Kazakhstan, and Belarus. NCI promotes the development of commercial activities in ten formerly closed cities in Russia that developed nuclear weapons. DOE also heads up the initiatives for Russia to turn HEU from dismantled weapons into civilian reactor fuel, and to accelerate the conversion of research reactors using HEU to using LEU. We are funding programs to eliminate the production of plutonium in Russia, aiming to replace Russia's three plutonium-producing reactors that provide energy for domestic consumption with fossil-fueled generating capacity and to shut down the reactors by 2007. Plutonium disposition activities, described below, also have received continued support.

The Future of Cooperative Threat Reduction Initiatives

While strongly supporting the CTR programs in Russia, we must recognize the challenges of working in and with Russia. These are very complex, difficult programs. On occasion, the General Accounting Office, an investigative arm of the U.S. Congress, has indicted the programs for "poor accounting practices." This just does not do the programs justice or recognize the delicate and tough nature of CTR programs. Without a doubt, progress on these programs is vital for our national security. The benefit to "our national interest" far outweighs the costs that some critics deride as "foreign aid." While frustrated with the Russians over their reluctant willingness to do their part regarding NCI programs, on balance the programs are worth the efforts. I have concerns over issues regarding needed "access" to Russian facilities. Current laws and practices make it extremely difficult for U.S. entities to enter the formerly closed, secret cities. However, we must realize this is a tough issue for the Russian Federation, and we must try to find innovative solutions. I strongly endorse the IPP program, which, to date, has obtained better results.

Expanding Our Nonproliferation Horizons

In 1996, expanding the focus of our nonproliferation activities as a response to warnings about biological and chemical weapons, Senators Nunn and Lugar and I sponsored legislation, the "Defense against Weapons of Mass Destruction Act," also known as the Nunn–Lugar–Domenici Act. It added funds for programs such as the one to help Russia to dispose of spent fuel from its nuclear submarines and to replace nuclear plants generating weapons-useable plutonium. It also established a new program to combat the use of biological or chemical weapons in terrorist attacks against the United States, setting up the initial first responders program in the United States that ultimately helped mitigate the disaster of the September 11 attack. After the terrorist attack on New York City, the city's fire department chief testified to Congress on the importance of first responders' training in minimizing the chaos from the World Trade Center attacks.

As a result of the Nunn–Lugar–Domenici Act, the MPC&A programs were significantly expanded. This act recognized the fundamental connectivity between earlier attempts to secure the dangerous materials at their source and the reality that local law enforcement and health officials must be trained to deal with the consequences should those materials get into the wrong hands.

After several years of work on expanded nonproliferation activities with Russia, the Clinton administration was determined to take a critical look at them. In typical Washington style, a task force was convened by the DOE, bringing together high-level, former government advisors under the bipartisan leadership of former Senate Majority Leader Howard Baker and former White House counsel to President Carter, Lloyd Cutler. The task force report, released on January 10, 2001, concluded: "The most urgent unmet national security threat to the United States today is the danger that weapons of mass destruction or weapon-useable material in Russia could be stolen and sold to terrorists and hostile nation-states and used against American troops abroad or citizens at home."[8]

The Baker–Cutler report recognized the necessity of bringing all nuclear materials around the world under control, and to that end proposed a $30 billion, ten-year program to dramatically increase the scope of activities in Russia and other FSU countries. This funding level would have increased the current spending level fourfold. With the change in administration, however, the momentum was lost and, frankly, factions in the incoming Bush administration who were hostile to Nunn–Lugar programs managed to stall things under the guise of a National Security Council review of such programs. After the terrorist attacks on September 11, executive branch resistance evaporated and ultimately Nunn–Lugar programs received their largest funding increases since the enactment of the Nunn–Lugar–Domenici Act of 1996. Today, the imperative to meet the terrorism threat on a global level is universally recognized.

Plutonium Disposition Programs in Russia

Background

I was very concerned that, following the 1994 agreements to reduce the U.S. and Russian nuclear arsenals, little significant action had been taken to address the plutonium issue beyond studies and evaluations. While the National Academy of Sciences had strongly recommended an important effort to dispose of significant quantities of excess weapons plutonium, progress in this area was slow; the subject was drifting.

In the spring of 1998, Representative (now Republican senator from South Carolina) Lindsay Graham and I cochaired a study at the Center for Strategic and International Studies (CSIS) in Washington, D.C., regarding the "Disposition of Surplus Weapons Plutonium." In March we rolled out the study's conclusions in an effort to urge the Clinton administration to more actively support a dual-track (MOX fuel and immobilization) approach for plutonium disposition—not just the disposal or "immobilization only" route, which we feared that some personalities with influence on the Clinton administration would prefer. The report endorsed the then-official U.S. government position that we should pursue a two-track strategy: burning surplus plutonium as MOX fuel in civilian reactors as well as vitrifying it and burying it as waste in Yucca Mountain.

It was our view that the MOX fuel option was essential for many reasons, most importantly because Russia viewed plutonium as an asset and not material to be stored as waste. The Russians have openly viewed immobilization as little more than an elaborate storage regime that would not serve to degrade the weapons potential of the plutonium involved. Our fear was that if the United States pursued immobilization alone, we would lose Russia's participation in the program.

Building on the CSIS report, I proposed the "internationalization" of a disposition program to more rapidly burn weapons-grade plutonium. Here was a case where I was too far ahead of what was feasible. As discussed further below, our idea, in summary, was:

- The United States and Russia would both ship their weapons-grade plutonium to Europe for MOX fuel fabrication.
- European MOX fuel capabilities would be used for MOX fuel fabrication in Europe.
- European MOX fuel based on the use of European reactor-grade plutonium would be stored, at U.S. expense, while MOX fuel with weapons-grade plutonium would be burned in its place in Europe, Russia, and the United States.
- MOX fuel derived from weapons-grade material would be burned in reactors in various countries with the caveat that only reactors under international safeguards would be used and those reactors must be in nuclear weapons states that already have weapons-grade plutonium.

Addressing the Eighth Annual International Arms Control Conference in April 1998, I said that the United States and Russia should work with the key industrialized nations—the seven most industrialized and democratic nations—to aggressively handle post–Cold War era surplus weapons plutonium in this way. I concurred with the conference report's conclusion that efforts must be accelerated to promote Russian endeavors to secure its plutonium stockpile and convert this material into a form unusable for weapons.

These ideas had been developing since my Harvard speech in October 1997. I also came to the conclusion that a more active dialogue with the Russians themselves was necessary. Congratulating Professor Adamov upon his appointment as minister of Minatom, I took the opportunity in a letter dated March 13, 1998, to ask about his interest in meeting me and my staff after the speech I was scheduled to give in Obninsk in July. I wrote:

There are two areas that would benefit from the earliest consideration. The first concerns how we might cooperatively accelerate the dismantlement of nuclear weapons, the conversion of classified fissile materials into forms suitable for use as valuable resources in commercial nuclear power programs, and the implementation of materials accounting and safeguards for sensitive nuclear materials. The second concerns the HEU Agreement, notably the recent difficulties in obtaining fair and prompt payment to Russia for the natural uranium component of LEU from Russian HEU under our government-to-government agreement of February 1993.

The Turning Point: My Trip to Europe and Russia in 1998

Senators Fred Thompson and Rod Grams and I went to France, Russia, and Germany in the summer of 1998 with, among other matters, the purpose of discussing options for the rapid destruction of excess Russian weapons-grade plutonium. While in France, we toured the French nuclear fuel facilities at La Hague in Cherbourg, visited the MELOX plant in Avignon, and traveled to Paris for meetings with industry, Cogema, and Commissariat à L'Enérgie Atomique (CEA) officials. We championed two of my ideas, labeled "Eurofab" and "Euroburn," for handling excess weapons plutonium. I suggested that the European capabilities for MOX fuel fabrication might be tapped in the Eurofab approach and that the extensive European experience with the use of MOX fuels might enable France to burn some of the MOX prepared from excess weapons plutonium.

At that time, these ideas were not accepted by the French. They were concerned with upsetting plans to use their own separated reactor-grade plutonium as MOX fuel in their reactors. There was also great concern about the logistics of introducing weapons-grade materials into their fuel cycle. The important point is that while these ideas were not accepted then, the concept of using French MOX fuel fabrication technology in a supporting role for the program ultimately was accepted. (French MOX fuel technology is now the basis for the U.S. and Russian MOX fuel fabrication plants.) As I proposed with Eurofab, the first test assemblies destined to be irradiated in the United States will be made at Cadarache, France.

With the French rejection of the Eurofab and Euroburn proposals, my staff and I got back onto the U.S. government plane thinking that we had wasted time and wondered whether to continue on to Russia. But after what my staff had gone through to prepare for the trip to Russia, we carried on. Setting up this trip was a nightmare for my staff. There were many exchanges between my staff, Minatom, and the Russian Embassy as we tried to gain permission for the United States Air Force (USAF) to land at the airport in Sarov, also previously known as the closed city Arzamas-16, as would be common practice for traveling U.S. senators. We were constantly told that the USAF could not possibly land there, and that the only airport that could accommodate a USAF jet was at Niszhny Novgorod, a three-hour car ride away. We were also told that the Russians would provide armed escorts from Niszhny Novgorod because "pirates" were sometimes encountered on that road. This possibility was not in the least appetizing as both the possible pirates and the armed escorts painted a chilling picture.

The Russians then suggested we take a Russian charter for the trip from Moscow to Sarov. While we continued to push, arguing that U.S. senators should only fly on USAF planes, our Air Force informed us fairly late in the process that they refused to land at Sarov because they had not inspected the runway! With that piece of information, we gave up the American military plane option and searched for a Russian charter.

We settled on Clintondale Aviation, an airline that was used by many U.S. oil companies doing business within Russia. We flew on a YAK-40, a small turboprop plane outfitted with landing gear that looked like it could support a B-52. Upon bouncing to a landing at Sarov, we immediately understood their insistence that our sleek USAF C-20 jet, which had carried us flawlessly from Washington to Moscow, could not possibly have landed at Sarov. We also quickly grasped why the Russian plane had such stupendous landing gear. The field was a mess, with jagged bits of old asphalt held in place by healthy grass and weeds.

We met with Minister Adamov at the Kurchatov Institute upon our arrival in Russia. Later, we traveled to Sarov, south of Moscow. I was among the first group of U.S. senators to enter a closed Russian city. There I met the director of the Russian Federal Research Center-All Russian Research Institute of Experimental Physics, Rady I. Ilkaev. It was fascinating to discuss his genuine interest in retreating from the Cold War mentality of these cities and to seriously explore commercial opportunities to shift his workers away from weapons work. I was most impressed with him and with his willingness to recognize the immense changes taking place in the world along with the important new opportunities for his laboratory.

From my first meeting with the director, and on the many other occasions I have had to meet him on his trips to New Mexico and Washington, our chats have ranged from comparisons of our respective grandchildren's accomplishments to the serious problems facing our two nations. I continue to be impressed with his vision to move his institution beyond the Cold War and to encourage commercialization of the vast technical talents in the closed cities. He and I have discussed improvements in our nonproliferation programs. Some of my enthusiasm for the IPP program derives from seeing the concrete progress that Dr. Ilkaev made at Sarov. He and I share the view that the world is a safer place if Russian weapons scientists have good paying jobs in the commercial sector, rather than going unpaid by a declining Russian weapons complex. In that latter case, hunger and concern for their families could potentially drive them to find it necessary to sell their skills to any of the rogue nations and terrorists all too eager to enlist them.

Dr. Ilkaev took me to visit the museum at Sarov where I had the opportunity to examine mockups of Russian weapons, including a mammoth sixty-megaton bomb. It was an intensely sobering experience to be standing, together with the bomb's designers, next to that weapon whose only purpose was to destroy our way of life.

In the museum I also viewed a mockup of a highly advanced miniature particle accelerator (the "radio-frequency quadrupole accelerator"), the inventor of which, Dr. Teplyakov, accompanied us. Ironically, he knew that his design had formed the basis for a new generation of accelerators in the United States. His pride of invention was very evident, but there was a touch of sadness as he noted that while he had designed the device from a theoretical perspective, there was no capability in Russia to machine the highly complex structure. He was proud that U.S. scientists at Los Alamos had taken his invention and shown it to work. He clearly wished he could have both invented the concept and been able to see it work in Russia.

I also recall another most interesting moment in that same museum. As I was leaving, I heard someone calling my name. The callers were two women from Los Alamos, New Mexico! You can imagine my surprise wondering how those two had possibly found their way into that remote corner of the world. As we talked, I learned that they were in Sarov as part of an exchange program between the two nations. They were both nurses, and were there to assist in improving health care in that closed city. This brought to me again how sharply the world had changed since the end of the Cold War. Since that time, I've watched with great interest the news on the continuing exchanges between the medical community in Los Alamos and that in Sarov. These exchanges have included work on improving care for diabetics in Sarov along with many other projects. Since that time, on many occasions, town officials, high school students, and medical personnel from Sarov have been in Los Alamos. Through these exchanges and projects, more bridges are built between our nations, and more Russian citizens come to better understand and feel more at home with the nation that they had regarded with so much terror for so many years. That improved understanding on both our parts is helping to build, I hope, a remarkable new era of increased cooperation between our nations.

Returning to Obninsk, where Russia's and the world's first civilian nuclear reactor began operation in 1953, I gave a speech to a youth conference. Dr. Pete Lyons still remembers me standing there in Obninsk on the Fourth of July, the day I usually give speeches back home in New Mexico. Instead, I was in Russia, at a convocation of Russian youth, on the subject of the global status of plutonium. I spoke to the young Russian men and women about the importance of controlling the various threats from plutonium and that effective disposition of weapons-grade plutonium would play a major part in realizing the promise of nuclear energy. This group was not large but it showed me that the Russians were very, very much aware of the potential for nuclear energy. Of course they are not scared of nuclear power! In fact, Russia favors efforts to build additional fast reactors while the United States has been stuck in the quagmire of indecision and only recently has been able to begin putting a new nuclear R&D vision together.

Returning to Moscow after the Obninsk speech, traveling at record speed in a Russian government motorcade down the center lane of a three-lane highway, we continued plutonium discussions with Deputy Minister Lev Ryabev. The talk was heady and fascinating, but my staff feared I would never make it back safely to do anything about our fruitful discussions because of the dangerous, harrowing road trip!

During our discussions in Russia, it was apparent that there was some difference of opinion among the leaders of Russia regarding plutonium disposition. However, Minister Adamov seemed to accept my suggestions to develop mutual agreements leading to rapid dismantlement of such excess weapons, conversion of classified shapes, and storage and disposition of the resultant excess weapons-grade material under international safeguards at Mayak. It was encouraging that the Russians appeared to be prepared to convert at least a significant fraction of their excess weapons plutonium to a more proliferation-resistant form through irradiation.

Even though at the time the French had dismissed our idea, by the time we left Moscow on our way to Frankfurt, we were incredibly enthusiastic about the idea of a U.S.–Russian bilateral agreement on weapons plutonium disposition. Once in Germany, my staff and I toured the Siemens MOX fuel fabrication plant in Hanau and investigated German interest in the plutonium disposition ideas. I'll never forget visiting with the head of the Siemens plant, who had tears in his eyes because the plant would never operate due to the political opposition within Germany. Regrettably, Germany was not to play a role in the unfolding plutonium disposition deal.

In addition to the revelation that key Russian leaders wanted to solve the plutonium problem, the visit to Sarov brought home to me the hard-core reality of the incredible infrastructure problems faced in the former Russian nuclear weapons cities and the immense challenge of encouraging commercial enterprises in these remote, closed cities. While our hosts expressed the most sincere of interests in commercial ventures, and support for the IPP, their prospects for success were obviously impacted by their remote location, extremely onerous access requirements, and an antiquated and somewhat unpredictable tax system. Just think, with a tax rate on a business at about 60 percent, it is clear what hurdles exist and how it is almost impossible for anyone to start up a business.

Access was extremely difficult, even for our congressional delegation. As I have already described, permission for our USAF plane to land was denied and the delegation and I were forced to travel on a Russian charter service. Long lead times for approval to visit were required. Security precautions seemed extreme—we were told to leave any cameras or recording devices on the plane before being allowed to deplane. Through my discussions with many workers, it was revealed that wages had not been paid for months. I learned that the nuclear cities' isolation and poor tax policies only added to their difficulties. At Sarov, a senior official remarked: "We have lost hope. We must rely on ourselves now." The director of Chelyabinsk-70, another closed city, committed suicide because of the difficulties there.

However, on a more positive note, we were briefed on the MPC&A program, and excellent examples of its successes were shown. Briefings on the IPP program also showed some successes. We were extremely interested in the ability of the NCI program to really change the situation in these closed communities, but also left with great skepticism that any modest program can successfully address the massive problems that were evident in Sarov.

The 1998 Moscow Summit Agreement: Plutonium Disposition Gets Underway

Immediately upon our return to the United States, I called and then personally met with President Clinton on July 27, relaying to him my concept for a U.S.–Russian bilateral agreement on plutonium disposition and how enthusiastic the Russian leaders had been. I argued for a U.S.–Russian bilateral agreement that would accomplish the following:

- Both countries should agree to establish a system of converting the surplus plutonium to another form that would be very difficult to reuse in weapons. Both sides would agree that the converted plutonium be stored and maintained under international guidelines, with a 10 MT per year rate for these steps.
- Both countries should begin developing a capacity to convert LEU and plutonium into MOX fuel for reuse as fuel in existing nuclear power plants in the United States, Russia, and possibly other countries.
- Both countries should agree to enter into a serious dialogue on the future dismantlement of U.S. and Russian nuclear weapons in a manner that would be fair to both sides.

I knew I had to seize the opportunity with the president as the Moscow Summit was looming just over the August horizon. As mentioned earlier, I sent Alex Flint to Moscow before the summit to participate in discussions to help ensure that the protocol's language was just right. In the joint statement, both countries agreed to remove in stages approximately 50 MT of plutonium from their weapons programs and convert the material so that it could never again be used for weapons purposes. Among the significant principles established by the agreement was that the goal could be accomplished either through consumption of the plutonium in current reactors or by immobilization in glass or ceramic form.

I accompanied President Clinton to the summit and was present at the momentous occasion when, with great hoopla, the "Joint Statement on Plutonium Disposition" was signed by Presidents Clinton and Yeltsin on September 2, 1998. It was a thrilling moment when President Clinton mentioned the senator that helped—that was me! I hailed it as a significant development, saying: "The Russians have agreed to enter into an enforceable agreement to take fifty metric tons of excess plutonium that could be used in 6,000 to 8,000 bombs, and convert it so that it cannot be used in bombs again. This is a very big decision for the world."

Returning to Washington, the real work began. Feeling that the State Department was not moving quickly enough on getting the program underway, we decided to put money on the table to get people's attention. At the same time that Congress appropriated the $325 million for the HEU deal in the omnibus budget package, we also managed to get $200 million included to jump-start Russia's stalled effort to convert excess weapons-grade plutonium to more benign MOX reactor fuel. We obtained a total of $525 million in 10 minutes from the appropriations committee chairman, Senator Ted Stevens, because the issue was of such moment. On the other side of the aisle, Democrats like Representative Obey were thoroughly impressed with the program concept and the urgent need to save the HEU Agreement. We solved two problems, in one swoop, in one afternoon! With regard to the plutonium funding, I should point out that it was very unusual for Congress to set aside funding for a program that did not yet formally exist.

After eighteen arduous months of negotiations on the details to implement the original 1998 bargain, the United States and Russia finally concluded in September 2000 the "Plutonium Management and Disposition Agreement." This bilateral agreement reduced the amount of plutonium that Presidents Clinton and Yeltsin had originally agreed would be disposed, from 50 to 34 MT. It required the construction of new facilities in the United States and Russia to fabricate the plutonium into MOX fuel, and allowed the United States to immobilize a portion of its plutonium rather than burn it in civilian reactors. Obviously disappointed in the reduction of plutonium, I was also worried that the agreement might be interpreted as setting a precedent for equal amounts of plutonium in future disposition initiatives, for the simple reason that estimates of excess Russian plutonium stocks are far larger than those of the United States.

There was progress. To meet the goal of each country converting plutonium into MOX fuel, DOE chose Savannah River as the site for the MOX fuel fabrication facility and Los Alamos for lead test assembly (LTA) fabrication. In March 1999, DOE had contracted with Duke, Cogema, and Stone and Webster (DCS) to build a MOX fuel fabrication facility at Savannah River. Later DOE changed course and decided to manufacture the LTAs at the Cadarache facility in France, thus returning, in part, full circle to our original idea of involving European fabrication capability.

Saving Plutonium Disposition Programs

After the elation of obtaining the bilateral plutonium agreement, it appeared that the new Bush administration was not convinced that the programs to assist Russia deal with its "loose nukes" problem were getting the intended results. Understandably, when a new president takes office, the new administration wisely undertakes a review of key national security projects. However, when President Bush took office and initiated a thorough review of U.S. nuclear policy, the timing and concurrent publicity of the National Security Council's report were especially unfortunate in their impact on plutonium disposition. The negative attitudes of some staff members of the National Security Council created an environment of immense confusion about the administration's plans, certainly in Congress and also with many other parties. Some of the members of the National Security Council staff questioned whether simply storing our plutonium might be preferable and less costly. They also wondered whether Russia might rather embark on an advanced reactor program that might ultimately provide a better disposition opportunity for weapons-grade plutonium. However, this would have been a very time-consuming and uncertain option. This kind of approach by the United States would have jeopardized all we had worked to establish over the years.

On November 13, 2001, in a bipartisan show of support for plutonium disposition programs in Russia, twelve other senators and I wrote to President Bush asking him to continue strong U.S. support for the Russian programs despite his National Security Council's negative review. Not willing to give up hard-won ground, we argued that:

> We must make every effort to guarantee that nuclear weapons materials in Russia are secure and permanently eliminated. While there are certainly a number

of important nonproliferation programs that are ongoing between the United States and Russia, it would be hard to pick one that is more important to global security than plutonium disposition or one that is more important to keep on track given the difficulty of getting it started. The risks to world stability from the theft or diversion of surplus plutonium by terrorist states are incalculable. The events of September 11 only reinforce this conclusion.

Within two months, not only did I receive a guarantee of the president's support for the programs, he even added more money! On January 9, 2002, President George W. Bush wrote to me:

> I agree with you that the United States and Russia should significantly expand cooperative efforts to reduce the threat of proliferation from the former Soviet Union. My Administration has completed a thorough review of our nonproliferation and threat reduction assistance programs with Russia. The review found that most of these programs are working well, are focused on priority objectives, and should continue. The review also found that several programs to prevent proliferation of weapons of mass destruction, sensitive materials, and expertise should be expanded. I will ask Congress for an overall increase in funding to support this vital mission.

The United States' and Russia's intentions to dismantle nuclear weapons were cemented further when Presidents Bush and Putin signed a pact at the May 2002 Moscow Summit to reduce weapons arsenals to the agreed-upon range of 1,700 to 2,200 warheads. During a trip immediately thereafter to Moscow, to take part in an international conference on nonproliferation, I was awarded an international humanitarian award for my work on nonproliferation issues. I participated in the destruction of a Russian nuclear warhead missile silo and took part in the dialogue about the necessary next steps in our quest to bring nuclear weapons under control. I am very grateful to the inspired leadership shown by my colleagues, Senator Richard Lugar and Senator Sam Nunn, whose efforts were essential to allowing me to accomplish many of my nonproliferation goals.

Plutonium Disposition Programs in the Beginning of the Twenty-first Century

The slow pace of the plutonium disposition program is admittedly frustrating, especially so when the change in administration led to an entire year's disruption of activities while the National Security Council conducted its review. Momentum in the program is crucial and we must maintain our goal of parallel progress in the United States and Russia on plutonium disposition. Continuing bipartisan support from the administration and Congress is absolutely essential to keeping this key element of our nonproliferation program moving forward.

The Russian MOX fuel program conceptually has strong and significant international participation, but it will be necessary to nail down the necessary funding, estimated to be in the range of $1 billion. To date, Russia has not committed any specific funds, but this is expected to change. The agreement to use the same French facility design at Savannah River and in Russia should facilitate progress. However, the United States and Russia have failed to agree on the necessary related scientific and technical terms for implementing the MOX fuel programs. The U.S. government, at the insistence of the State Department, is insisting as of early 2004 that Russia provide the United States with broad protections against liability, similar to that provided under the CTR program.

The impasse between the United States and Russia over the liability issue is adversely affecting the pace of the disposition program. Specifically, it has become necessary to delay the start of the construction of the Russian and U.S. MOX fuel fabrication plants by ten months (as of the time of writing) from the summer of 2004 to May 2005. Nevertheless, the DOE continues to maintain it can meet its commitment to begin MOX fuel fabrication in 2008.

Senator Richard Lugar and I met with Minister Rumyantsev on November 4, 2003, to try to identify ways to move forward on the liability issue, as well as with the IPP and NCI programs. Both sides are interested in resolving the liability problem, but it needs to be given more high-level attention and greater urgency. We should be open to looking at various options. At a Senate Budget Committee hearing on the fiscal year 2005 funding request for the Department of State on February 26, 2004, I pressed Secretary of State Colin Powell to "find the right people" within the State Department to solve the dispute that is jeopardizing progress in the plutonium disposition program.

The liability impasse has become a matter of great personal disappointment to me. I can only say "Shame on both of our nations" when we allow an issue like liability indemnification to stop progress toward disposition of excess weapons plutonium. Our priorities are not set straight. This program is not just in the interest of the United States. It is not just in the interest of Russia. It is a concrete step for all countries toward a safer world. This begs for leadership from the governments of both the United States and Russia. I believe that we can overcome this stumbling block with Colin Powell's firm leadership.

Emerging Proliferation Threat

I have immensely enjoyed the challenges of working on proliferation issues. However, the problem is that those of us working on such issues have directed the lion's share of our attention on Russia, when there obviously are several other critical problems in nations such as North Korea and Iran. Senators Nunn and Lugar and I have underscored the importance of the problem and raised the concern that not enough attention is being paid in the Senate on these emerging proliferation threats. The problems of proliferation from Russia worry me but they don't worry me nearly as much as the reality that some other countries evidently are developing nuclear bombs.

The world is waking up to the challenge to current nonproliferation policies and institutions that are posed by other nations, including Iran and North Korea. Iran we now know has been clandestinely pursuing, under the cloak of legitimate commercial nuclear power activities, a wide range of nuclear activities, including construction of centrifuge cascades for enrichment. We worry that a state with military intentions could pursue development of nuclear weapons under the guise of nuclear programs for peaceful purposes and pull out of the NPT after having received nuclear technology for peaceful uses, without any meaningful consequences to such a decision.

North Korea has pursued a clandestine nuclear weapons program outside the purview of the IAEA. It is assumed that North Korea purchased enrichment technology from Pakistan; it has been operating a reprocessing facility at Yongbyon and has been in the process of building three graphite reactors. In 1994, under the so-called Framework Agreement with the United States, it agreed to close its graphite reactors and the reprocessing facility in exchange for two 1,000 MW LWRs, security assurances, and oil supplies from a consortium including the United States, Japan, and South Korea. When confronted with evidence that it had not ceased efforts to produce nuclear bomb materials in October 2002, North Korea admitted at the time to pursuing a secret enrichment program, withdrew from the NPT, expelled the IAEA inspectors, and announced intentions to continue its nuclear weapons development program. At first acknowledging its illicit enrichment program, the North Koreans have since changed their tune and disavowed the evidence.

North Korea's behavior confirms all my worst fears about proliferation risks. As this book was in press, we heard even more ominous reports about this rogue nation's nuclear program. Dr. Abdul Qadeer Khan, the Pakistani scientist who has admitted selling nuclear secrets throughout the world, claims to have actually seen the North Korean secret underground plant where the North Koreans have three nuclear devices.[9]

Not only do the North Koreans brag about having nuclear weapons, and having reprocessed eight thousand spent nuclear fuel rods, but they have evicted international nuclear weapons inspectors. Finally, the country's leaders say that they have restarted a reactor at Yongbyon, a claim confirmed by some U.S. intelligence officials.

This has been the classic proliferation disaster: A nation agrees to end its nuclear program. It secretly continues the programs, amassing spent nuclear fuel rods. Then, and this is the critical part, it gains access to the specialized knowledge necessary for making a functional weapon. At some moment, it then announces that it has the weapon and challenges anyone to do anything about it.

Some of North Korea's Asian neighbors have tried to hide from the realities of the situation. Some have claimed that the North Koreans were merely bragging, in order to keep clout in the world. Most others said that it was impossible for the North Koreans to really have created a weapon that would work. The recent revelations, however, put even more pressure on Asia's other countries to join with the rest of the world in a unified effort to end the North Korean program. Remember that the final segment of the proliferation nightmare comes when stateless terrorists gain control of a nuclear weapon from a rogue nation. That could be the final stage of the North Korean irresponsibility.

It has also been asserted that Iran purchased enrichment technology from Pakistan and has been experimenting with reprocessing techniques since the mid-1980s. The recent confession of Pakistan's Abdul Khan, while useful if it represents a stopping point, does nothing to undo the problems he has caused. His assistance to Libya, North Korea, and Iran has done terrible damage.

The issue over Iran's intentions has been made all the more urgent because Russia is helping Iran complete the Bushehr reactors, which the United States fears will provide further cover under the auspices of NPT-sanctioned civilian nuclear commerce for a weapons programs. The IAEA passed a resolution on November 26, 2003, deploring Iran's failures to live up to its nonproliferation obligations. The United Kingdom, France, and Germany have tentatively offered Iran technology, equipment, and fuel cycle services in return for its pledge to give up sensitive fuel cycle activities and its agreement to sign an additional "protocol" with the IAEA that will significantly broaden the agency's safeguarding opportunities in that country. It is not certain whether Iran has made the decision at its highest political level to renounce its nuclear intentions.

Following the gruesome events on September 11, 2001, it became clear that U.S. efforts in the nonproliferation area and in countering terrorism needed to be ramped up. To this end, Senators Joseph Biden and Richard Lugar and I introduced in May the Nuclear Nonproliferation Act of 2002, also called the Domenici–Biden–Lugar Act. It was proposed as a stand-alone bill with ten bipartisan cosponsors and later passed as an amendment to the Defense Authorization Act of 2003. Most of the provisions survived the conference process and were placed in the final fiscal year 2003 Defense Authorization Bill passed by Congress and signed into law by President Bush.

Passage of that bill was an extension of my efforts over a six-year period to help create a more comprehensive U.S. nonproliferation policy, above all else recognizing the global character of proliferation and terrorism challenges. The first responders program I launched in 1996 was reauthorized. And the act broadened most of the DOE nonproliferation programs to include an international focus, not just a focus on the FSU. It did this in a way to encourage cooperation with the FSU wherever possible and recognized the vital role to be played by the IAEA. Among key provisions, it created a new initiative to fund collaboration with Russia to study problems associated with dirty bombs and to offer assistance to other countries facing this challenge. It extended MPC&A assistance to countries beyond Russia and authorized a program for international cooperation on the design of proliferation-resistant nuclear energy technologies.

We have had some recent successes in better managing weapons-useable materials. In 2002, the NTI, in cooperation with the U.S. government, the IAEA, and Yugoslavia, undertook a yearlong effort to remove 100 pounds of weapons-grade uranium, enough for three nuclear warheads, from a poorly guarded facility in Yugoslavia. The material was removed in 2002 from a nuclear reactor in the Vinca Institute of Nuclear Sciences in Belgrade, Yugoslavia, and transported to a safer facility in Dmitrovgrad, Russia.

The United States provided funding for yet another covert, successful attempt to bring highly enriched material from a sensitive location. In September 2003, with the help of Romanians, Russians, the IAEA, and Americans, eight canisters of approximately 30 pounds of HEU were taken from the research reactor site of the Pitesti Institute for Nuclear Research west of Bucharest, Romania, and loaded onto a Russian IL-76 cargo plane and taken to a site in Siberia. The operative plan is for the Russians to blend down the HEU at the Novosibirsk Chemical Concentrates Plant, owned by Minatom, and sell it. The United States will pay the costs for converting the research reactor in Romania to low enriched fuel.

A second major piece of the puzzle in dealing with emerging threats is to account for and secure vast amounts of materials that could be used in radiological dispersion devices or dirty bombs. Senators Biden and Lugar and I cosponsored the "Nuclear and Radiological Terrorism Threat Reduction Act of 2002" to strengthen efforts through our Department of State. Our proposal featured the creation of repositories to provide temporary storage of radioactive sources in Russia that could be stolen by terrorists, specialized training for overseas "first responders" to handle radioactive emergencies, and inspection of cargoes at foreign ports of embarkation rather than on U.S. shores.

I also have proposed the establishment of a "fast reaction analysis team" made up of weapons of mass destruction experts, equipped with radiation detectors and isotope identifiers, to quickly identify and seize radioactive materials around the world. In our war against terrorism, we must support other countries' abilities to detect, interdict, and then prosecute traffickers of nuclear, radioactive, and dual-use materials. Nuclear materials must be kept under very tight control and the programs that we started in Russia must be broadened to enable international cooperation. Global cooperation is essential to prevent global terrorism.

The third piece of the puzzle is to ensure the strength of our domestic infrastructure against the threat of terrorist attack. I can well understand the concern among the general public about whether the security measures at nuclear power plants are sufficient to deter or withstand a terrorist attack. The gist of the concern is whether a determined attack, such as those that destroyed the World Trade Center as well as a wing of the Pentagon, would result in a release of radiation that would create great damage to the public and the environment. I think the concern is understandable and certainly worth addressing, as the NRC and the nuclear industry have done.

But the public is not well aware that their concerns have already been addressed to a very great extent. Nuclear power plants are protected by formidable security measures, making them some of the most secure industrial facilities in the country. Our security posture has only improved since the September 11 attacks. These measures include enhanced fortification of perimeters, additional barriers to protect against vehicles, and additional surveillance equipment, among others. While the industry acted independently to enhance security measures following the 2001 terrorist attacks, the NRC made many of these measures mandatory in early 2002. Finally, addressing a specific and valid concern, it is reported that computer modeling has shown nuclear power plant containment structures can withstand impact by a large commercial jet with maximum plausible force without a release of radiation to the environment.

A New Paradigm for U.S. Nonproliferation Policy

I have repeatedly emphasized the need to make nonproliferation progress in both military and civilian areas of nuclear technologies. It also is clear that nuclear power will only be able to realize its full potential if governments and the public have confidence that the use of technology will not significantly aggravate risks of proliferation.

There is a complex interplay between the military and civilian aspects of nuclear technologies. On the one hand, nuclear technologies offer immense benefits to modern society. But on the other hand, if we fail to control the potential military applications of nuclear technologies, we may never be able to fully benefit from their contributions to mankind.

The good news is that we have made huge strides over the past several years and are poised to take advantage of political and technical solutions to the remaining concerns. Thanks to contributions by the United States and several other countries, numerous concrete steps have been taken to shore up the nonproliferation regime. Just to mention a few of the biggest improvements in Russia as well as around the world that I have witnessed during my tenure in the Senate:

- The United States and Russia have entered into the Moscow Treaty, which requires a decrease in each of our countries' arsenals of strategic nuclear warheads to between 1,700 and 2,200 by the year 2012.
- The Bilateral Plutonium Production Reactor Agreement requires the shutdown of fourteen U.S. and ten Russian plutonium production reactors. We have signed an agreement that will lead to the shutdown of the last three such reactors in Russia.
- Russia will convert 500 MT of highly enriched, weapons-grade uranium (equivalent to 20,000 bombs) into low enriched fuel for U.S. nuclear power reactors; 201.5 tons have been converted as of the end of 2003, representing material for 8,059 nuclear warheads, through the Megatons to Megawatts program mentioned above.[10]
- In 1994, the United States identified 174 tons of excess HEU that would either be blended down for use in the civilian nuclear sector or disposed of as waste.
- Both the United States and Russia will each dispose of 34 MT of plutonium.
- As of the beginning of 2004, the excess fissile materials removed from Russian and U.S. stockpiles are the equivalent of 30,000 eliminated nuclear weapons.
- CTR programs have led to the elimination of 5,809 nuclear weapons, 1,212 ballistic and cruise missiles, 795 missile launchers, 92 long-range bombers, and 21 ballistic missile submarines.[11]
- Three nations—Ukraine, Belarus, and Kazakhstan—are now free of nuclear weapons because of CTR initiatives. When the Soviet Union collapsed, Kazakhstan had 104 intercontinental ballistic missiles, 1,040 nuclear warheads, and 40 strategic bombers, as well as the nuclear testing site at Semipalatinsk where 400 tests took place. Through the $100 million provided under the auspices of the Nunn–Lugar initiated CTR program, all weapons were removed from that country by 1995 and the testing site was finally destroyed by 2000.

- While the situation in the closed Russian cities was especially dire in the period from 1992 to 1999, the situation is much improved now, thanks largely to the Russians themselves, but also because of the very good cooperation from the United States.
- South Africa abandoned its nuclear weapons program.
- Brazil and Argentina foreswore their military nuclear programs.
- Libya has declared its weapons development work to the IAEA and is in the process of destroying these programs under the watchful eye of the IAEA.
- Within the past year and a half, the IAEA and the Bush administration have launched major initiatives to locate and secure radiological materials.

We must build on decades of U.S. efforts to protect our nation and the world from the threats of nuclear holocaust should nuclear materials fall into the hands of terrorists or rogue nations. Many actions need to be taken to further reduce the threats posed by the possible flow of materials of mass destruction into the wrong hands. This fact was made tragically clear on September 11, 2001.

We must make removal of HEU from research reactors a priority. There are over 130 research reactors in 40 countries using HEU.[12] This material is a serious threat because of its vulnerability to theft and its usefulness in making a weapon of mass destruction. Consequently it should be removed from its current locations and replaced with low enriched material. Congress must appropriate the necessary money to assist this effort. Everyone in the nuclear industry has a stake in the success of this endeavor because of the negative impact a bomb in the hands of a terrorist would have on the industry.

On a global scale, my vision of nuclear power includes the deployment of reactors in developing countries to bring the benefits of electricity, clean water, and perhaps, in the future, hydrogen to these nations. As noted elsewhere in this book, the greatest increase in the demand for electricity is likely to occur in the developing world. However, to date there have been various obstacles to introducing nuclear power in such countries. The grids in several such countries have been too small, and their technical capabilities too limited to absorb large-scale nuclear units.

Accordingly, I support looking at new options to facilitate introducing nuclear reactors in developing countries. These could include the potential use and application of small modular reactors with attractive characteristics for remote communities that otherwise must rely on shipments of relatively expensive and environmentally undesirable fuels for their electric power. To be acceptable, such reactors would have to be inherently safe and relatively cost effective, contain intrinsic design features that would deter sabotage or diversion, require infrequent refueling, and be largely factory constructed and deliverable to remote sites. To this end, I have supported funding the design work for a plant to demonstrate the viability of such small modular reactors.

Variations of the new Generation IV reactors will be important in achieving this goal with their focus on passive safety, improved proliferation resistance, and less waste generation. New plant designs are extremely promising, such as the Pebble Bed Modular Reactor under consideration in South Africa. That reactor has many of the features of a Generation IV reactor, including passive safety, and its modular design lends itself very well to rapid construction. Its size, at 110 MW could provide all the electricity for smaller countries or could be combined with multiple units to supply larger needs.

As an example of such a system, Toshiba is currently developing the "4S—super, safe, small, and simple" reactor, with assistance from U.S. national laboratories. Major features of the concept under development include a fast neutron, sodium-cooled reactor with a core lifetime of thirty years, meaning that it would not be refueled. It could range in size from 10 MWe to 50 MWe. The reactor and steam generator could be embedded underground with the turbine building located above ground. The core would be fueled with a uranium-zirconium mixture. Reflectors would move up the assembly rods slowly over the thirty-year period, but would fall down in the case of an accident to make the assemblies subcritical. The reflectors would be driven by electromagnetic impulsive force. Current promotional literature on the 4S reactor state that its cost would be $1,500 per kWh. Toshiba hopes that the first 4S reactor can be commercialized for a site in Alaska, with construction beginning in 2008, testing by 2010, and operational in 2011, only three years after the start of construction.

The Navy has designed and received NRC peer-review approval of the design for a new Virginia-class reactor. While not directly applicable to the concept of the "black box" reactor I am suggesting, the Navy's new reactor is not that far from where we need to go. A highlight of this reactor's design is that it will not require refueling for thirty-three years! I believe that the United States can help develop the technology that will lead to the development of a small, passively safe, affordable reactor for export to the developing world.

There are many reasons why the international nonproliferation regime must be reinforced and strengthened by advancements in technology. The United States must continually push for those improvements so that nuclear programs already started in countries do not provide cover for obtaining nuclear weapons or an easy source for a terrorist to obtain some type of radioactive material.

The spread of nuclear technology and ever-changing geopolitical realities force us to move beyond institutional barriers to erect technical barriers to proliferation and terrorism. Advanced reactors and associated fuel cycle systems themselves must provide additional proliferation barriers.

We should move to advanced, closed fuel cycle systems in order to conserve uranium resources, provide for better waste management, and decrease the proliferation potential inherent in an open fuel cycle or one that depends on Purex reprocessing (which is described in detail in chapter 9.) The additional technical barriers to proliferation can be built into these advanced systems. This is what Gen IV is all about.

Technical improvements must be complemented by reasonable incentives for countries to forego indigenous fuel cycle capabilities as I outlined in my speech on the fiftieth anniversary of President Eisenhower's "Atoms for Peace" speech. This concept was recently supported by President George W. Bush in his speech before the National Defense University in Washington, D.C.[13] President Bush suggested exploring concepts in which nuclear fuel supply would be guaranteed in return for a country's agreement not to build reprocessing or enrichment operations. I support such a proposal if done under the auspices of the IAEA control.[14] Countries that already have nuclear weapons and fuel cycle capabilities could together offer lifetime-guaranteed fuel cycle services to countries that only need nuclear power plants for electricity.

Our best and brightest minds are surely up to the task of proposing, for example, new, realistic models based on competition between "industry consortia" rather than between fuel cycle suppliers. Such an industry consortium could possibly include a supplier of each aspect of the fuel cycle, with companies from several different countries, thereby providing assurance of services in return for a country foregoing development of sensitive technology operations. The Iranian case is worth noting. Three countries have tentatively offered to cooperate with Iran in providing fuel cycle services if it acts responsibly. At this time, the United States cannot participate due to a lack of diplomatic relations with Iran. In the future, it would be in the United States' best interests to consider participating in such arrangements both for the sake of our nuclear suppliers and because the United States would require adherence to the most stringent nonproliferation standards.

In summary, we must recognize the linkage between management of all nuclear materials on a global scale and the future of commercial nuclear power. We have programs to help control the Russian weapons materials, but they could be making much faster progress than they are. The threat of international terrorism should lead us, our Russian partners, and others to rapidly expand efforts to secure global military- and civilian-sector materials. In the spirit of fostering progress in making the dream of nuclear power come true while adhering to the goals and norms of our nonproliferation regime, these specific recommendations should be considered:

- provide the moral, technological, and policy leadership in the nonproliferation arena by pursuing the most advanced nuclear R&D program in the world, such as those in the Advanced Fuel Cycle Initiative (AFCI) and Gen IV initiatives, and ensuring it has a strong nuclear infrastructure to be able to make the necessary advances in proliferation-resistant reactors and fuel cycles;
- executive branch appointment of a person with oversight over all U.S. nonproliferation programs;
- Congress must continue efforts and provide adequate funding for our nonproliferation program with Russia to upgrade the protection of all sites in Russia where nuclear weapons are stored and to ensure that all stocks of excess weapons materials in Russia are converted to non-weapons-useable forms as soon as possible;
- expand the scope of current cooperative threat programs such as MPC&A worldwide, especially into Pakistan and India;
- Congress must appropriate funds to allow an acceleration and expansion of programs in Russia for HEU blend-down;
- continue support for the Moscow-based International Science and Technology Center, which aims to place former Soviet weapons scientists in nonweapons work;
- accelerate the conversion of reactors using HEU to the use of LEU fuels through increased U.S. financial incentives;
- foster global cooperation in containing fissile materials as suggested in the 2002 Domenici–Biden–Lugar Nonproliferation Act;
- address nuclear and radiological (dirty bombs) terrorism, specifically standards and rules for the security of potential radiological sources and support the president's Proliferation Security Initiative as well as those undertaken by the IAEA; provide resources to localities to respond to radiological emergency making it a less desirable avenue for terrorists;
- support strengthened, universal application of IAEA safeguards in non-weapons states through widespread acceptance of the Additional Protocol;
- assure the IAEA has robust financial, personnel, technical, and political resources to monitor countries' NPT obligations and compliance;
- insist that the U.N. Security Council take action against NPT signatory countries who have breached their obligations;
- without undermining the NPT, develop strategies to prevent countries from pursuing nuclear weapons under cover of the NPT or from exiting the NPT with the allowed ninety-day notice once weapons capabilities are accumulated while a member of the NPT;

- support a comprehensive upgrade of physical security requirements in each and every country that utilizes nuclear materials and improvements in each country's regulatory regimes;
- minimizing plutonium stocks in civilian reprocessing fuel cycles and in the military sector;
- contain the spread of proliferation made possible through enrichment or reprocessing technologies by improving IAEA detection capabilities as envisioned under the Additional Protocol, developing advanced fuel cycle technologies that do not allow for separation of weapons-useable materials, and exploring innovative institutional arrangements and incentives for nonweapons states to forego fuel cycle steps such as enrichment and reprocessing;
- pursue technology and policy advancements that limit or eliminate access to special nuclear materials, specifically by minimizing production of plutonium isotopes that are weapons-useable, eliminating the need for enrichment facilities, reducing access to fissile materials, eliminating streams of separated plutonium, and minimizing the transport of weapons-useable materials; and
- continue emphasis on the control of military weapons materials.

It is an irony of history that the commercial nuclear power industry, child of the nuclear weapons program, may ultimately play a key role in destroying nuclear weapons materials. Whether it be by burning weapons-useable HEU in civilian nuclear power plants à la the Megatons to Megawatts program, burning much more excess plutonium in existing reactors, designing advanced reactor systems that do not separate out weapons-useable materials, developing advanced safeguards and physical materials controls technologies that reduce the possibility of proliferation, or prodding the institutions responsible for nonproliferation to adopt the most stringent of rules and consequences regarding proliferation—the civilian nuclear sector can provide the leadership and wherewithal to help meet the challenge.

Our long-term goal should be a world without nuclear weapons, but that goal will only be achieved after many, many years of patient progress toward intermediate nonproliferation goals. Each step along this journey must remain focused on further reductions in the availability of materials for weapons or terrorists' purposes. Let us stay the course, expand initiatives on a global scale, and pursue the reduction of conflict worldwide through the improvement of the economies in developing countries.

9

The Waste Disposal Conundrum

Our nation must maintain nuclear energy as a viable energy source far into the future. With advanced technologies, it can provide electricity for centuries. It is a clean and reliable source of baseload power that will be essential in powering our economic growth for future generations, just as it is already a vital component of today's economic successes. However, for nuclear energy to continue to be viable, we must make progress on the path toward the disposal of nuclear wastes. There is no denying that these wastes represent an area of risk; but every energy source requires a balance of benefits and risks. The risks associated with nuclear waste are ones that we can fully control.

Dealing with nuclear wastes is perhaps the most frustrating challenge for the future use of nuclear energy because of the politicization of our decision-making process. However, it is solvable. I believe that the barriers to progress in opening our first repository are entirely political, not technical. While we must continue to make progress on opening the geologic waste disposal facility at Yucca Mountain, Nevada—and I am whole-heartedly supporting the full funding of the program—it is not clear to me that the current plan for the direct disposal of spent nuclear fuel is in the ultimate interest of America. Further, depending on our future power demand and electricity options, we may someday need to recover the tremendous energy that remains in spent nuclear fuel. No matter what our future course of action may be, a geologic repository will be needed for the disposal of some of the residual wastes.

We have demonstrated the feasibility of the storage and disposal of nuclear waste in my state of New Mexico. In March 1999, the Waste Isolation Pilot Plant (WIPP) opened near Carlsbad. WIPP is the nation's first repository for the permanent disposal of defense-generated radioactive waste from the research and production of nuclear weapons. Now is the time to make the same progress on the high-level waste front and remove the perceived "waste management barrier" to the expanded use of nuclear energy in America.

What Is Nuclear Waste?

There are several types and forms of waste from the commercial nuclear fuel cycle, commercial applications of nuclear technologies, and the production of nuclear weapons, each with their own particular type of management and disposition requirements. In the case of defense-related wastes, plutonium and other isotopes were separated from production reactor and naval propulsion reactor spent nuclear fuel.[1] There are also plutonium-contaminated wastes from weapons-fabrication plants. Commercial wastes are produced by the nuclear reactors that generate electricity, facilities that process reactor fuels, and by a host of other institutions and industries that use nuclear materials for civilian purposes.

Most of the radioactive wastes, whether commercial- or defense-generated, are labeled as: high-level waste, including spent fuel; low-level waste; uranium mill tailings; mixed waste; and transuranic wastes. In the current debate about the future of nuclear power, we are primarily concerned with the progress on the disposal of the high-level wastes. To help understand the issues better, here's a very short primer on several categories of nuclear wastes.

High-level wastes comprise either the wastes generated from the chemical processing of spent nuclear fuel or the spent nuclear fuel itself.[2] The United States plans to solidify high-level wastes from reprocessing, mostly generated from the nuclear weapons program, and dispose of them in a repository. Spent nuclear fuel, on the other hand, is discharged from a utility's nuclear reactor, cooled in storage pools at the reactor site, sometimes stored in air-cooled metal or concrete storage casks, and then transported to a U.S. government-owned geologic repository for final burial, when available. While the U.S. Congress established a 10,000-year period for evaluation of repository performance, spent nuclear fuel should be isolated from the environment for approximately 300,000 years primarily because of the radiotoxicity of the actinides in the fuel, plutonium, neptunium, americium, and curium.

Transuranic waste is generated from many operations (other than the reprocessing of spent nuclear fuel) that involve plutonium, including the fabrication of nuclear weapons from plutonium.[3] It consists of clothing, tools, rags, residues, debris, and other such items contaminated with small amounts of radioactive elements (mostly plutonium). Transuranic elements are radioactive, man-made, and have atomic numbers greater than uranium; hence the term "transuranic"—beyond uranium. Even though the level of radioactivity is mostly low, transuranic wastes remain toxic for thousands of years, as is also the case with high-level wastes, and therefore require very long-term management. Transuranic wastes generated from our defense programs are now being buried in the WIPP facility in Carlsbad, New Mexico.

Low-level waste (LLW) refers to most wastes that are not classified as spent nuclear fuel, high-level, transuranic, or mill tailings. It comes from a myriad of sources including power plants, hospitals, laboratories, and industrial plants. Generally these wastes are disposed of in shallow land burial plots, and most require little shielding during transportation or disposal because of their low level of radioactivity.

Mixed wastes contain both hazardous chemicals and radioactive components. Disposal policies are complicated and under development because there is greater uncertainty about the biological effects of chemicals than there is uncertainty about radiation effects.[4]

Uranium mill tailings are the remains from the mining and extraction of the natural uranium from the ore found in the earth. While the tailings are not generally hazardous, their disposal is carefully controlled because of the radon gas emitted.

To put the nuclear waste issue into its proper context, in the industrialized countries with nuclear power programs, radioactive wastes comprise less than 1 percent of the industrial toxic wastes generated. During the past fifty years, the U.S. nuclear industry has produced 10,000 times less nuclear waste than a year's worth of toxic chemical production in our country.[5] High-level wastes, which are, in turn, only a tiny fraction of the radioactive wastes, are the primary concern. However, one meter of concrete is sufficient to stop the radiation emitted from high-level wastes.[6] As I learned during my 1998 visit to La Hague, in the French reprocessing system, a single 150-liter glass canister contains the waste (fission products and actinides) from 360,000 families of four heating their homes with electricity for one year.[7]

The life-cycle impact of nuclear energy—including all emissions from the construction of plants, mining of the uranium, production of nuclear fuel, generating plant operations, disposal of the wastes, and decommissioning—is among the lowest of any form of electricity generation, including hydropower, coal, natural gas, biomass, wind, and solar. Pound for pound, uranium-235 yields a million times more energy and a million times less waste, per kilowatt-hour produced, than wood, coal, oil, or gas.[8]

The nuclear power industry generates many types of nuclear wastes that must be properly managed and eventually disposed of. The major area of concern is primarily about the high-level wastes, as management and burial of the other forms of waste are already proceeding forthwith in the United States, generally without any negative consequences to the public's health and safety.

U.S. Waste-Management Policy

Before President Carter deferred the reprocessing of spent fuel in April 1977, a commercial reprocessing plant operated in the United States from 1966 to 1972. The president's deferral resulted in the United States effectively losing its flexibility in the long-term management of spent nuclear fuel. This decision, based on a false premise that our allies would follow our example and halt reprocessing as well, forced us into the decision to directly dispose of spent nuclear fuel in an underground repository.

The decision taken by President Carter's administration, codified in Presidential Directive #13, was based on the assumption that an "open fuel cycle" would pose less of a proliferation risk than the "closed fuel cycle" because the reprocessing technology used at that time resulted in the extraction of plutonium (the terms "open" and "closed" fuel cycles refer, respectively, to direct disposal of spent nuclear fuel and reprocessing of spent nuclear fuel with disposal of the processed materials). It was argued that such separated plutonium could be used to make an explodable nuclear device, albeit one that might not have as great a certainty of exploding as one made from high-quality weapons-grade plutonium. Under this reasoning some policymakers concluded that the spent nuclear fuel should not be reprocessed so as not to leave stockpiles of separated plutonium that could theoretically be made into a weapon. President Carter thought that if the United States took the "high road" and did not reprocess, the world would follow and be safer for it.

Well, how wrong he was. France, Japan, the UK, and Russia continue their reprocessing programs, fabricate the resulting plutonium into MOX fuel, and France burns it in their LWRs. Their stockpiles of separated plutonium have been safely stored at various sites and have not proven to be sources of proliferation. Moreover, history has shown that other countries prefer to obtain weapons-usable materials by enriching uranium rather than reprocessing spent nuclear fuel. North Korea appears to be the only known exception to the enrichment route to weapons fuel. President Carter was wrong on two fronts—no one followed our lead and proliferation mainly flowed from the pursuit of uranium enrichment technology.

While the Reagan administration revoked the deferment of reprocessing and plutonium recycling in 1981, the U.S. utility and nuclear industries' interests in those options had waned in the meantime. This was due partially to concerns over the economics of recycling and also to the prospective nuclear waste legislation. The Nuclear Waste Policy Act of 1982 (NWPA), as discussed below, provided for a framework for making decisions to dispose of high-level wastes in geologic formations. President Carter's decision set in motion the development of a waste disposal strategy that was based on the direct disposal of spent nuclear fuel. Within five years, Congress passed a law establishing our disposition strategy.

The NWPA created a process and a set of milestones through which the DOE would select and characterize potential sites for geologic repositories, recommend sites to the president for two repositories, and begin development of the first repository. However, the plan for siting two repositories was not to be.

Originally, former Congressman Mo Udall—Democrat from Arizona, chairman of the House Interior Committee, and a leader in environmental issues—brokered a deal that ensured the building of two repositories: the first in the western section of the country and the second in the east after the western repository began operation. In 1986, DOE selected three sites in three western states for site characterization—Texas, Washington, and Nevada. The preliminary cost estimate was $800 million per site. While the funding for the waste repository activities comes from a surcharge on generated nuclear electricity, both the Reagan administration and Congress began to balk at the total price tag for the repositories.

The investigation for a second site was focused on granite formations in seventeen mostly eastern states, including North Carolina. However, when in June 1986 Senator John P. East, Republican of North Carolina, died while in office, the Republican governor, Jim Martin, appointed Congressman Jim Broyhill to fulfill the remaining term. Congressman Broyhill had been the ranking member of the House Energy and Commerce Committee.

The location of the waste repositories then became a contentious issue in the 1986 election in North Carolina (as well as in New Hampshire and several other key states). The Reagan administration decided in the fall of 1986 to cancel the investigation of the second-round repository. Many pundits speculated at the time that the announcement was politically motivated to help Broyhill in North Carolina. Ironically, in the end, Broyhill lost to former governor Terry Sanford by 56,000 votes out of 1.6 million cast. After the 1986 election, Democrats regained control of the Senate.

On the day of the cancellation announcement by President Reagan, K. P. Lau, my energy aide, was at the executive office of the White House negotiating with the Office of Management and Budget on the cost portion of the comprehensive uranium bill and received the advance notice of the pending announcement. The meeting was a follow-up to an earlier meeting on budget issues between Don Regan, chief of staff of the White House, Jim Miller, Office of Management and Budget director, and me. I had requested support from the White House on the comprehensive uranium bill and it just so happened that the same afternoon, K. P. Lau had scheduled a private meeting for me to meet with Mo Udall to solicit his support for the legislation as well.

By the time we got to his office, Congressman Udall had learned about the announcement and was fuming. He felt betrayed by the cancellation of the second repository. He did not mince words about his displeasure and summed up his feeling with one of his famous jokes, thereby diffusing a tense moment. Thus, the second repository siting activities came to an ignominious end.

The NWPA established the Office of Civilian Radioactive Waste Management within DOE to carry out the federal waste management program. The NWPA also provided that DOE contract with utilities for acceptance and disposal of their spent fuel and with others to dispose of civilian spent nuclear fuel and high-level radioactive waste beginning no later than January 31, 1998.

The nation's nuclear electricity generators are required by law to pay for all the disposal costs through payment into a nuclear waste fund of a fee of one-tenth of 1 cent for every kilowatt-hour (1 mil per kilowatt-hour) of nuclear-generated electricity. Electricity generators have already paid over $19 billion in principal and interest into this fund.

Congress redirected the waste program through the Nuclear Waste Policy Amendments Act of 1987 (NWPAA) in order to stem growing opposition to the program and to try to curtail rising costs of screening three potential repository sites. The NWPAA directed the DOE to stop work at all potential sites except Yucca Mountain, Nevada, located on a tract of U.S. government-owned land about 100 miles northwest of Las Vegas. The DOE was directed to determine if the proposed Yucca Mountain site would be suitable for a repository, and if so, seek congressional authorization to begin repository construction. Subsequently, the 1988 Draft Mission Plan Amendment announced a five-year delay in the start of waste acceptance from 1998 to 2003.

The NWPAA contained provisions to establish the Office of the Nuclear Waste Negotiator with the intent that the negotiator would locate a state or Indian tribe to volunteer a site for a monitored retrievable storage facility—that is, an interim storage facility. In January 1995, the Office of the Nuclear Waste Negotiator was closed without successfully siting such a facility.

In November 1989, the secretary of energy submitted a comprehensive reassessment of the civilian high-level radioactive waste program to Congress in response to congressional inquiries regarding program schedules, program management, and contractor integration. The high-level waste management program schedule was revised, further delaying the projected start of repository operations this time from 2003 to 2010.

On February 14, 2002, Secretary of Energy Spencer Abraham forwarded to the president his recommendation of the Yucca Mountain site for development of a nuclear waste repository. On the following day, President George W. Bush notified the Congress that he considered the Yucca Mountain site to be qualified to move on to the next step in the process to develop a geologic repository.

The governor of Nevada vetoed the selected site, as was his right under the NWPA. In accordance with this law, the Nevada governor and legislature had sixty calendar days to issue a "notice of disapproval" of the Yucca Mountain site. Governor Guinn issued a notice of disapproval to the U.S. Congress on April 8, 2002.

As a result of this notice, Congress was required to vote to approve the repository-siting designation. Congress had ninety days of continuous session to approve a Resolution of Repository Siting. The House of Representatives voted on May 8, 2002, to approve the resolution by a vote of 307 to 117. The Senate voted on July 9, 2002, approving the resolution in a vote of 60 to 39. On July 23, 2002, President Bush signed the Yucca Mountain siting resolution into law.

The DOE is now in the process of developing a license application that it expects to submit to the NRC by the end of 2004. The DOE further expects to obtain construction authorization in 2008 and then an updated license application to receive and possess waste and operate the repository by 2010, pending the outcome of the lawsuits brought by the state of Nevada and others.

The world scientific community's consensus view is that all types of nuclear waste may be safely buried in a deep geologic repository in a manner that protects the health and safety of people and the environment. Several countries have now decided to proceed with geologic disposal, including the United States, Finland, and Sweden.

Waste Program Lessons Learned from France

I was impressed by so much of what I saw in France during my visit in 1998. They have made an immense commitment to nuclear power, and at the same time have taken strong steps to educate their public on the benefits and risks of nuclear technologies. They deserve great credit for the extent to which they have integrated their waste management operations into the community. At La Hague, I recall being impressed by a superb public education system, and the fact that their environmental monitoring system was fully open to the public. At La Hague, peaceful dairy cows grazed up to the boundaries of the facility. I've also never forgotten the moment when I walked out onto the concrete floor of their high-level waste storage area, and stood on top of the canisters of the vitrified high-level wastes.

One day on the Senate floor, I presented my colleagues with a version of the French glass waste canister, similar to the ones I saw at La Hague, which held the total amount of waste that a family of four would generate over a period of twenty years. You would need several dump trucks to deliver the comparable waste from this family of four using coal-fired electricity!

France is investigating geologic disposal in a clay formation at a site not far from Paris. With a substantial nuclear program of fifty-nine operating reactors producing about 50 GWe of electricity, or 80 percent of its electricity needs each year, France is designing a waste management program that reduces the amount of disposal space needed to the greatest extent possible. France has chosen the reprocessing and recycling route, reducing the amount of waste by a factor of four to five. Further recycling may reduce the waste volume by a factor of ten. In addition, France is pursuing a major transmutation research and development program that it hopes will lead to a reduction in waste volume up to a factor of one hundred. This could be accomplished by transmuting the minor actinides as well as the higher plutonium isotopes.

In the French program, after the spent nuclear fuel is stored and cooled in a pool at the reactor site for a period of six months to one year, the material is transported to Cogema's La Hague plant in the northern tip of the Cotentin Peninsula of France. When the spent nuclear fuel is received, it is remotely unloaded either under water or in a hot cell, and placed in baskets for interim storage for five to eight years. The assemblies are then removed and cut into pieces. The nuclear material is dissolved in nitric acid and the fission products are separated from the uranium and plutonium using chemical solvents. The uranium is concentrated in the form of uranyl nitrate and the plutonium is packaged in oxide form in sealed containers. The fission products and minor actinides are then prepared for final burial using the vitrification process.

The uranium and plutonium are then available for further usage. In the case of the uranium, it could be reenriched before being made into fuel again. In the case of plutonium, it is mixed usually with depleted uranium. France has been fabricating MOX fuel at its Melox facility on an industrial scale for more than ten years. The mixture contains 92 percent depleted uranium and 8 percent plutonium. After being recycled in a thermal reactor, the spent MOX nuclear fuel assembly will only contain 4 percent plutonium. As you can see, by recycling, a nuclear power plant in France consumes plutonium. Currently, the French electric utility company, Electricite de France, has plans to recycle plutonium gradually in most of its 900 MWe reactors by operating them with 30 percent MOX fuel.

As mentioned, the fission products, which make up between 3 to 5 percent of the spent nuclear fuel assembly, are vitrified and slated for burial. Ninety-four to 96 percent of the spent nuclear fuel (uranium) is recycled, along with the 1 percent plutonium. (The percentages differ depending upon the burn-up rate in the reactor.) Compared to direct disposal of the spent nuclear fuel, by reprocessing and recycling, the volume of waste to be disposed of under France's scenario is reduced by a factor of five (thus reducing costs) and the radiotoxicity of the waste is reduced by a factor of ten (thus reducing costs and potential health and environmental impacts).

It is worth noting that more than 18,000 MT heavy metal of spent nuclear fuel have been reprocessed at the La Hague plant. In excess of 60 MT of plutonium have already been recycled, more than 50 percent of that amount through burning the plutonium in MOX fuel in twenty of France's 900 MWe PWRs. It has also pursued R&D on partitioning and transmutation, which along with geologic disposal and waste conditioning and long-term storage, are the three areas of waste management research required by France's law on waste management passed in 1991.[9]

Investigating France's waste management program brought home to me that France's careful use of reprocessing technology presented options for waste management that we simply don't have in the United States. It helped to confirm my belief that we have to do better for the American public than simply dispose of spent fuel in Yucca Mountain—our waste management policies need significant rethinking.

When I returned from France, I vowed that I would help redirect our waste management program. We must not stand for the old tactic to stop nuclear power by insisting plants be closed down unless and until we can put the waste away forever. In America communities threaten riots if waste is moved into their area, when in France, there it safely and securely sits. We must learn lessons from France's nuclear program.

Waste Management Challenges

Delays in Licensing and Opening a Geologic Repository

As envisioned in the NWPA, the DOE originally planned to open the nation's first repository in the year 1998. In 1987, it admitted it would delay the repository's opening by five years, to the year 2003. Only two years later, in 1989, it further delayed the expected opening date to the year 2010. The repository is at least twelve years behind schedule; the license application is not expected until December 2004. Causing further concern, the DOE has not chosen an interim storage center and spent fuel is piling up at nuclear plant sites. The federal government defaulted on its expected commitment to begin transporting the current backlog of spent fuel from reactor sites to such an interim storage location in 1998.[10]

The NRC is required to approve all of DOE's waste disposal activities and license its facilities and transportation containers. I'm well aware that there were hundreds of outstanding issues identified by the NRC during their prelicensing deliberations for the repository. And the DOE is well aware that they must address each and every one of the NRC issues before the commission is going to move toward a final license. In many meetings with NRC commissioners, I've always been impressed with their intent to deal with this or any licensing issue through careful study of the relevant scientific facts. The NRC has the expertise to evaluate these outstanding issues, and I'm confident that they will do so with great care. It is not up to me, a U.S. senator, to decide on the complex scientific issues that will eventually determine the fate of a license for Yucca Mountain.

The EPA, responsible for developing site-specific standards for the Yucca Mountain site, has set out regulations for the amount of radiation acceptable for a person living close to a high-level waste disposal facility. The EPA ruled that this person may not be subjected to more than 15 millirems of radiation for the first 10,000 years after final waste disposal. This is twenty times less than the average annual radiation dose that a person receives from natural, everyday radiation sources. The EPA further established a 4-millirems water standard.

The National Academy of Sciences analyzed the EPA's recommended standards prior to issuance and determined that it had not provided a technical rationale for its groundwater protection approach.[11] There were also concerns that the EPA had not performed health-benefit and cost analyses of its regulations. But the EPA regulations are symptomatic of a larger problem of basing regulations on the LNT model (see chapter 5). Issuing regulations protecting the public at very low levels, and at levels below which radiation effects have yet to be proven, is a very costly business. Given the questions surrounding the validity of basing regulations on the LNT approach, we may have painted ourselves into a box without health or economic justification.

I opposed the decision to allow the EPA rather than the NRC to set standards for the disposal site. The NRC has a history of acting as a nonpolitical organization, in sharp contrast to the political nature that the EPA has shown in the past. We need unbiased technical knowledge in setting the radiation protection standards—this is no place for politics. As I noted, even the National Academy of Sciences was critical of EPA's proposed radiation protection standards for Yucca Mountain. However, the decision has been made and the NRC must now enforce the standards through its regulation of the repository.

Further contributing to potential delays in licensing the first repository, the state of Nevada (as well as Las Vegas, Clark County, and the Natural Resources Defense Council) is engaged in a full-scale legal battle with the federal government over the Yucca Mountain repository. Hiring big-gun lawyers and other support, Nevada has filed at least six discrete actions against the program and regulatory actions of the agencies developing the programs. These cases have been consolidated and are now awaiting action before the Court of Appeals in Washington, D.C., one court away from the Supreme Court. Decisions are expected in 2004. The essence of Nevada's challenge, in the words of Nevada's Nuclear Waste Project Office, is:

- [U]nder the U.S. Constitution, 49 states may not gang up on a single, politically isolated state to impose on it an unwanted burden without a compelling rational basis. With the abandonment of any geologic isolation criteria for the site, such a basis no longer exists for Yucca.
- [W]hen the Energy Department changed its site suitability rules in 2001 and abandoned geologic requirements for the Yucca site, it violated the Nuclear Waste Policy Act, which requires geologic isolation of waste.
- [T]he Energy Department committed procedural violations of the National Environmental Policy Act and the DOE failed to evaluate the impacts of the project in accordance with that Act and the Nuclear Waste Policy Act.
- [T]he Energy Secretary's and President's site recommendation is legally void because the decisions were based on both flawed siting analyses and a flawed environmental impact statement.
- [T]he EPA did not follow the requirements of the Nuclear Waste Policy Act, the Energy Policy Act, and the Safe Drinking Water Act when promulgating the radiation release standards. The state argues that the EPA "gerrymandered" the Yucca site boundary to meet radiation release standards for drinking water dilution and limited the regulatory compliance period to a time that precedes the time of the known peak hazard from the repository.
- [T]he NRC's repository licensing rules violate the Nuclear Waste Policy Act and the Atomic Energy Act. This is argued because the NRC cut off the regulatory compliance time period prior to the time Nevadans will experience Yucca's peak radiation hazard; they set no minimum requirements for the geology or the geological fitness of the Yucca site; and they fail to require actual defense-in-depth through application of multiple barriers, i.e., natural geologic and man-made barriers.[12]

In the event Nevada is unable to stop the Yucca repository through the courts, then it is likely the state of Nevada will mount another full-fledged campaign replete with procedural, technical, and safety aspect challenges during the licensing hearings that the NRC will hold.[13]

I am very concerned that progress on licensing this first waste repository will be hamstrung by court challenges. We saw what happened in the 1980s when opponents of nuclear power hijacked the regulatory process by filing lawsuits. Delay after judicial delay increased the costs of nuclear power plants to the point where some utilities were forced to abandon projects just because the cost was becoming unacceptably great, not because the plant was not needed or was not safe. The resolution of valid concerns and issues belongs before the NRC as provided by law. However, I do not believe the courts are the proper venue to evaluate scientific and technical issues.

Failure to license, construct, and operate the nation's first waste repository, or to agree on an alternative management strategy, could have a devastating impact on the potential resurgence of the U.S. nuclear industry. Some states have hitched their nuclear policies directly to the opening of a repository. Although it may be repealed, Wisconsin state law requires that additional reactors cannot be built there unless the Public Service Commission finds that a federally licensed facility exists to dispose of the spent fuel. Demonstrable progress on this front is essential today. Without progress the public hue and cry will be: "What if Yucca never happens? Without a plan for the waste, America should not produce any more than we already have."

The Transportation Link

Transportation of nuclear waste is a red herring issue. I've been very sorry to see the overblown concerns on transportation being held up to the public by the antinuclear groups as a reason why nuclear power is not acceptable. Apparently the opponents of Yucca Mountain are so intent on winning this battle that they are willing to use transportation issues to frighten the American people into abandoning nuclear energy. That would be a colossal mistake for our nation and would seriously undermine national security.

The simple fact is that transportation of nuclear materials takes place on a daily basis and the risks associated with radioactive materials are safely managed. Radioactive-materials transport is also an operation that has been extensively studied and engineered for success. Packages for transporting spent nuclear fuel are designed to withstand severe transportation accidents. In addition, extensive analyses and field tests show that there is virtually nothing a saboteur could do to the shipping cask that would result in a significant risk to the public.

In the United States, as well as in other countries, the record for transporting spent fuel is superb. In Europe 70,000 MTHM (metric tons of heavy metal) of spent nuclear fuel have already been shipped through densely populated areas without incident. The U.S. Navy has shipped 783 containers of high-level waste over more than 1 million miles since 1957 without a harmful radioactive release. Since 1960, we have shipped spent nuclear fuel about 2,700 times for combined distances of over 1.6 million miles. Sure, there have been a few accidents involving commercial spent nuclear fuel—eight in total since the mid-1960s. But there has never been a release of the radioactive contents in the transport casks. When the Yucca Mountain facility is up and running, it is expected that it would receive 70,000 shipments of waste between 2010 and 2034 in 3,215 train deliveries (three shipments per train) and 1,079 truck deliveries. Opponents need to remember that the shipping casks for spent fuel are designed to withstand severe transport accident conditions, and routes will be carefully chosen to further limit risks.

Additional Repositories

There are 49,000 MT of spent nuclear fuel at reactor sites across America today. The current generation of nuclear plants discharges slightly more than 2,000 MT of fuel each year. Civilian nuclear power plants will have discharged 65,000 MT by 2010 and 70,000 MT by the year 2015, while the current statutory capacity at Yucca Mountain is 70,000 MT, of which 63,000 is reserved for civilian use. If Congress allows the statutory expansion of Yucca Mountain to 120,000 MT, it is likely it could handle the spent nuclear fuel for the lifetime of the current generation of reactors. Capacity might be increased by adding more drifts and tunnels, extending the available space by a factor of two or three, but that would require changes to current law.

Under the nuclear power growth scenario projected by MIT, which I wholeheartedly support, current spent nuclear fuel disposal plans would necessitate the construction in the future of more geologic storage facilities. With an increase in U.S. nuclear power capacity to 300 GWe by the year 2050, all of the existing power plants would produce 380,000 MT of spent fuel for storage by the year 2110, assuming sixty-year operating nuclear plant lifetimes and a discharge rate of 20 MT per gigawatt-year. Continuing nuclear growth after 2050 would lead to an even larger amount. Even assuming that the first repository could handle an expanded capacity of high-level waste, it is clear that there may be a need for more geologic disposal facilities.

The NWPA requires the secretary of energy to make a determination in the 2007–2010 time frame as to whether another geologic repository will be required. If we continue to generate electricity by nuclear power at the current rate, more repositories may be needed. If we accelerate the nuclear program in the United States to make hydrogen-powered fuel cells and to reduce greenhouse gasses in the atmosphere, we will need more repositories if the DOE directly disposes of spent nuclear fuel.

This challenge leads me to the conclusion that the United States must pursue research on alternative fuel cycles and on technologies such as reprocessing and transmutation that can lighten the repository burden. We need to do the research today that can allow tomorrow's leaders to decide whether these options can lead to economical and safe systems with better energy efficiency, reduced risks, and enhanced benefits.

Don't Bury a National Energy Treasure

The energy value of the spent nuclear fuel currently stored at nuclear power plants today is roughly the equivalent to that contained in 6 billion barrels of oil—equal to two years of American oil imports.[14] For example, spent nuclear fuel from PWRs commonly deployed in the United States and initially made from uranium enriched with 4 percent uranium-235 will contain approximately 1 percent plutonium. If this plutonium is directly disposed as is the current U.S. plan, it will account for nearly all of the radiotoxicity of the waste a few centuries later. Worse yet, this plutonium, and as a matter of fact, the residual enriched uranium that makes up 95 percent of the spent nuclear fuel, represents a huge amount of potential energy.[15] Just think, complete fission of 1 gram of plutonium or 100 grams of uranium produces more heat than complete combustion of 1 ton of oil.[16] In actuality, the United States has been safely burning plutonium in its reactors over the past thirty years since President Carter's decision. By the end of a four- to five-year cycle of burning nuclear fuel in a LWR, about 50 percent of the energy will have actually been produced by burning plutonium that is created by burning the enriched uranium fuel.

If the renaissance of nuclear power comes to pass, as I know it will, we will need significant increases in uranium resources worldwide. As explained in greater detail in chapter 6, I have reached the conclusion that concerns regarding the availability of such resources require us to revisit our plans for the direct disposal of spent nuclear fuel and consider advanced alternatives to the current fuel cycle. The uranium and plutonium available in our current and future spent fuel inventories is indeed a national treasure and significantly contributes to our diverse and therefore more secure supply of energy. Energy resources should not be needlessly wasted.

A final concern is that if spent nuclear fuel is directly placed into a geologic repository, it could become a source of plutonium for potential terrorists or governments bent on developing nuclear weapons. After a period of about fifty to one hundred years, the high activity fission products, which present a barrier at first to anyone wanting to handle or steal the spent nuclear fuel, decay. At that point, the proliferation resistance of the open fuel cycle is negated. It is time to recognize that the once-through fuel cycle is not more proliferation resistant than the closed fuel cycle, especially under the scenarios being researched by our government today.

Recommendations

I sense a pervasive lack of public confidence in the United States regarding our waste disposal program, perhaps because of continuous program schedule revisions, but equally so because of the loud obstructionism with which certain groups frequently bombard the public. The folks opposed to nuclear power have held up the missteps and delays in our waste disposal program as primary reasons to oppose the licensing of Yucca Mountain. They attack the fairness of the siting process. They challenge the technical merits of engineering systems planned for the repository. They disagree that the long-term performance of a repository can be accurately projected. Their entire approach has been destructive rather than constructive.

The politicization of the nuclear waste disposal issue has hindered sensible public debate regarding the truths and risks of disposing nuclear materials. I agree with Richard Rhodes who said, "Nuclear waste disposal is a political problem in the United States because of widespread nuclear fear disproportionate to the reality of relative risk, but it is not an engineering problem, as advanced projects in France, Sweden and Japan demonstrate."[17]

While I support progress in developing an underground nuclear waste repository at Yucca Mountain, I believe that our single-minded focus on the permanent disposition of spent nuclear fuel rods does not serve our nation well overall. It is simply not obvious that permanent disposal of spent fuel is in the best interests of our nation. It's even less obvious to me that we should equate the terms "spent fuel" and "waste."

Since Yucca Mountain may not be able to accommodate all the spent fuel from our current generation of nuclear plants as well as the DOE's spent nuclear fuel and high-level waste, we clearly either need a better solution or more repositories. Given the lack of local public support enjoyed by Yucca Mountain, I don't think any of us should relish the prospect of siting more Yucca Mountains. As mentioned above, this decision regarding a second repository is required by law in the 2007–2010 time frame.

Depending on future demand and options for electricity supply, the United States may need to recover the tremendous energy that remains in spent nuclear fuel. At the current rate of consumption of uranium resources, and especially under an expanded nuclear scenario, known resources of uranium are likely to be exhausted by the end of the century. Expansion of current mines and development of new sources cannot be relied upon to provide a certain supply of uranium for nuclear reactors.

In summary, the United States must rapidly move ahead with R&D of next-generation fuel cycles that generate far less waste and extract the full energy benefit from each gram of fuel. Since this is a long-term effort, in order to be successful it needs strong research programs today. Following are a few recommendations that could help set us on the right path of nuclear waste disposal.

Develop a Centralized Interim Storage Program

Our nation will be far better served if we begin moving spent fuel from individual reactor sites to a single, well-secured repository. It is folly to leave it stored in temporary facilities at 131 sites in 39 states across America. It would be prudent to plan for centralized safe storage of the United States' spent nuclear fuel.

Preference aside, however, it is important to point out to the American public that the safety and security of spent nuclear storage pools at these reactor sites is more than adequately assured by physical barriers, operating procedures, and security precautions. Since September 11, the utilities have taken additional measures to prevent, and if necessary, withstand a terrorist attack. There is no justification for costly measures to remove fuel rods from the storage pools and place them in dry storage casks. We do need, however, a more coherent strategy to deal with the buildup of utilities' fuel on site.

As a first step toward a more rational integrated waste-management strategy, I favor prompt development of centralized interim storage in a monitored, highly secure, fully retrievable configuration. This would reduce the pressures on our utilities' on-site spent nuclear fuel storage pools, allow the DOE to make good on its pledge to accept spent nuclear fuel, and give our national laboratories some breathing space to develop advanced spent nuclear fuel management strategies. Such monitored storage facilities could allow future generations of Americans to evaluate their own needs for energy and decide on an appropriate reuse of spent nuclear fuel or final disposition. Most importantly, we should not prejudice this decision today. In a very real sense, a centralized, monitored, retrievable storage facility for spent nuclear fuel would represent a national nuclear fuel reserve for future generations.

Pursue Advanced Waste Management Strategies

Pursuing advanced fuel cycle research and development would broaden our waste management options, reduce the need for more repositories, reduce the costs for disposal, provide our nation with the technology if it should decide to pursue a closed fuel cycle, and better utilize our fuel resources. Advanced fuel cycles can improve waste management via processing technologies by reducing the volume of waste; for example, by eliminating the largest percentage of its volume, uranium. Transmutation can eliminate most of the long-lived radioactive constituents (primarily the transuranic actinides), shortening the required isolation time, and reducing the need for additional repositories.

What is this potential new approach to waste management all about? Figure 9.1 demonstrates the basic concepts involved in reprocessing, recycling, burning, and transmuting spent nuclear fuel. There are many interrelated steps in the conceptual system.

To better understand such an approach, first let's review the composition of the spent nuclear fuel when it is removed from a reactor core. It is composed of roughly 95.6 percent uranium, 3 percent stable and short-lived fission products, 0.3 percent cesium and strontium, 0.1 percent iodine and technetium, 0.9 percent plutonium, and 0.1 percent other long-lived actinides.[18] The fission products are originally more radioactive than the actinides. Most of the fission products decay in less than a year. There are longer-lived fission products, such as cesium 137, strontium 90, and iodine 131, that must be isolated for a long period (between five hundred and one thousand years) so that they are no longer a hazard to the environment. Other fission products have extremely long half-lives of thousands of years, with radioactivity that may be so low as to not require such long periods of isolation from the environment.

The "bad actor" actinides of most concern are plutonium, neptunium, and americium. These are the components that lead to the need for isolation for thousands of years. As you can see, the spent nuclear fuel has many different components—and each could be treated in different ways for disposal.

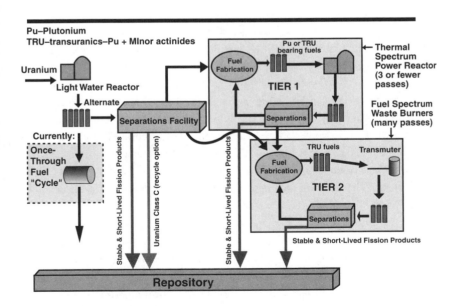

Figure 9.1.   Multitier Fuel Cycle Approach to Nuclear Waste Management. *Source: AAA Quarterly Report, April–June 2002, Los Alamos National Laboratory, LA-UR-02-5515, August 30, 2002.*

Under an advanced waste-management strategy, rather than sending the spent nuclear fuel directly to a geologic repository, it would first go to a separation processing facility. Everyday, American households separate their trash into recyclables and garbage; the recyclables are further separated into plastic and paper goods before being put in the proper container for disposal! This is the same concept being envisioned for the spent nuclear fuel. Some of the parts of spent nuclear fuel are garbage that go directly to the repository; other portions that are recyclable, such as plutonium, are processed again.

First, at the separation processing facility, the spent nuclear fuel would be broken down into different components. The United States has already demonstrated that uranium can be separated out of the spent nuclear fuel. It comes out of the process with a slight amount of contamination because no process is 100 percent efficient. It has been demonstrated (at laboratory scale) that we can make this separation at 99.99 percent efficiency, and perhaps higher. In fact, laboratory workers have actually held the separated uranium—from an experiment that was conducted with about a gallon of actual spent fuel from a power reactor—in their hands. The separation efficiency is very important because with the reduced contamination level, there is now the option to recycle the uranium by using it as feed material for enrichment and then reuse as reactor fuel. Or you can dispose of it as waste. Just taking out the uranium, which is over 95 percent of the volume of the spent nuclear fuel, is a huge boon for the waste disposal program.

The program's original goal was to reduce the contamination to a sufficient enough level for the waste to be classified as "Class C" or better. Class C waste, according to law, can be disposed of cheaply with shallow earth burial and does not need to go to Yucca Mountain. Bench-scale testing has demonstrated that not only can we achieve the Class C designation, but we can now look at the potential for recycle. In my mind the more recycling we can do, the better, because we are minimizing waste. The separated uranium would make a great feed material because it has a slightly better uranium-235 enrichment than natural uranium and therefore would reduce the overall cost of enrichment. There would also be significant savings due to reduced uranium mining requirements.

The significance of this is that separating out the uranium will reduce the volume of waste requiring a Yucca Mountain-type of disposal site by approximately an order of magnitude. *The U.S. would still require a geologic repository.* However, the cost of waste handling and emplacement would change dramatically. If we go to the point of purifying the uranium stream, it becomes the preferred source of materials for fuel reenrichment because it is already partially enriched above that of natural ore, thereby reducing enrichment costs and totally avoiding the mining and milling costs of new ore. These two attributes represent a significant cost savings. Separating the uranium stream from spent nuclear fuel is a significant first step toward assuring a more sustainable nuclear fuel cycle.

At this initial separation processing stage, the stable and short-lived fission products can be taken out. The fission fragments in the spent fuel that account for the most radioactivity and heat, strontium and cesium, can be sent to the repository after proper conditioning. Again using France as an example, all of the fission products and some of the actinides could be vitrified—in other words, mixed with molten glass—and held in a storage facility. France is storing its vitrified wastes until its repository is ready for their final disposition. Such glassified wastes are far more stable and less toxic than the spent nuclear fuel rods we now envision placing in a repository.

The second step in the separation process holds even greater promise for achieving sustainability. The second step involves removing the "bad actor" actinides I described already. By taking these out you can reduce the volume down to a small percentage of the original fuel volume. More importantly, you are removing the long-term toxicity problems and the heat-producing isotopes. The consequence is that Yucca Mountain becomes the only repository the nation would ever need and the performance confirmation period can shrink dramatically from the 10,000-year legislated limit of today down to only a couple of hundred years. This is because the remaining fission product waste will decay away to natural background in this much shorter period. The separated fissile materials from this second step, plutonium and neptunium, etc., will require management. However, it has already been demonstrated that they can be destroyed in reactors or accelerators (as described in greater detail further along in this chapter). Alternatively, they can be safely packaged and stored in aboveground vaults. This is a much more attractive proposition than repository disposal from both safety and security standpoints.

In the next stage, it can be envisioned that the plutonium would be recycled in existing or advanced LWRs. In the French system, the pure plutonium retrieved from reprocessing at the La Hague facility is then turned into its oxide form, mixed with depleted uranium oxides, and then burned in LWRs. The MOX fuel contains 8 percent plutonium. At the end of its lifespan, the recycled fuel only contains 4 percent plutonium—a net consumption of plutonium. In other words, burning the MOX fuel burns much of the plutonium that was created in burning the original uranium-enriched nuclear fuel assemblies.

Currently, the separation process being developed by the U.S.'s Advanced Fuel Cycle Initiative (AFCI) program, known as the uranium extraction (UREX) process, involves the separation of plutonium, other transuranics, and fission products from the uranium. The presence of minor actinides such as neptunium, americium, and curium, which are highly radioactive, render this separated plutonium waste stream unattractive to would-be proliferators. Experiments to date have been encouraging. The DOE is doing laboratory-scale demonstrations on the separation of plutonium and neptunium from the other actinides and fission products and plans a laboratory-scale separation of americium and curium and then cesium and strontium from the spent nuclear fuel.

It is important to note that the United States is pursuing a reprocessing technique different from that followed by the French at La Hague. The French process, known as PUREX, is derived from the process used in the United States for weapons work, and results in this separated stream of plutonium without the actinides. (The fact that PUREX results in a stream of pure plutonium was what drove the proliferation concerns of President Carter.) Research indicates that the UREX technology makes this plutonium/neptunium stream unattractive to potential proliferators.

The next phase involves transmutation, defined in *Webster's* dictionary as "the conversion of one element or nuclide into another either naturally or artificially." In the case of nuclear wastes, atoms in the waste stream would be irradiated by neutrons produced by either an accelerator or a nuclear reactor. The neutron would be absorbed, thereby producing new isotopes that might have shorter half-lives. The long-lived fission products can also be transmuted into safer elements through a process of neutron capture. For example, the long-lived fission products such as iodine 129 and technetium 99, which are major contributors to the long-term radiotoxicity of the spent nuclear fuel, could be separated out and then fabricated into a target/fuel, placed in a reactor, bombarded with neutrons, and destroyed—"transmuted." For example, when technetium 99, which has a half-life (the time it takes to lose half of its initial radioactivity) of 212,000 years, is bombarded with neutrons, it is transformed into technetium 100, which decays into stable and harmless ruthenium in a matter of just minutes.

To summarize, almost all that would remain in the waste stream would be short-lived fission products that can be incorporated into compact, durable waste forms and might even be kept in separate aboveground storage sites. If the fission products are placed in the repository, the waste density can be greatly increased by simply waiting for the short-lived fission products to decay, thus increasing the capacity of the repository.

Eventually, the residual plutonium and minor actinides could be sent to a separation processing facility to be processed into fuel for reuse in a special reactor, such as a fast reactor or an accelerator-driven system. At this stage, the actinides are fissioned again and destroyed. Unfortunately, it is not efficient to fission the actinides in a LWR. More promising are the fast reactors. These reactors can burn the actinides from the LWRs as well as those created in the fast reactors themselves. The actinides could also be destroyed in an accelerator-driven system. It may be possible to irradiate these actinides more safely using external neutron bombardment created by an accelerator because the system can be operated in a subcritical mode with larger accident margins. This is why research is needed—to determine the most efficient and cost effective way to fission the actinides.

Since the radiotoxicity from actinides remains significant for a period of time exceeding 10,000 years, by burning the actinides the necessary time for isolation of the actinides from the environment is significantly reduced. This increases confidence in the repository's ability to safeguard the public from radioactivity over time.

Figure 9.2 compares the toxicity of nuclear waste under a direct disposal scenario to a successful recycling, burning, and transmutation scenario. If the spent nuclear fuel were to be disposed of directly, then it would not reach the level of the toxicity of natural uranium for almost 300,000 years.

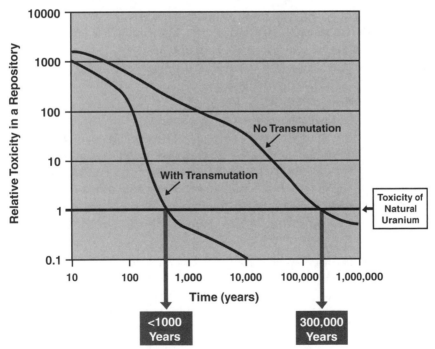

Figure 9.2.   Effective Toxicity Reduction due to Transmutation. *Source:* U.S. Department of Energy Office of Nuclear Energy, Science, and Technology, "Report to Congress on the Advanced Fuel Cycle Initiative: The Future Path for Advanced Spent Fuel Treatment and Transmutation Research," January 2003, figure III-1.

Under the alternative scenario, it reaches this level in less than 1,000 years.[19]

The bottom line is that by taking out the uranium (95.6 percent) left in the spent nuclear fuel, burning the plutonium (0.9 percent) as fuel in a light water or high-temperature gas reactor, transmuting the long-lived fission products (iodine and technetium, 0.1 percent), and fissioning the long-lived actinides (0.1 percent), the volume and radiotoxicity of wastes that must be emplaced in a geologic repository will be significantly reduced. This would reduce the cost of the repository and perhaps even eliminate the need for another repository.

In the words of Nobel Laureate Burton Richter, chair of the Subcommittee on Nuclear Waste Transmutation of the DOE's Nuclear Energy Research Advisory Committee: "Transmutation has the potential to reduce the required isolation time to on the order of a thousand years greatly reducing concerns about unlikely geologic events. It also has the potential to reduce the amount of material needing long-term storage, thereby reducing the size and cost of repositories."[20]

The DOE has laid out an R&D program that can, given the necessary funding, determine in five to ten years the viability of transmutation on an industrial scale. If successful, the government will then have to decide whether to proceed to a demonstration project. Should that demonstration in turn be successful and the system deployed, the amount of material requiring long-term isolation would be drastically reduced, the isolation time would also be drastically reduced, and the plutonium would be gone.[21]

Argonne National Laboratory has studied nine different transmutation systems, including seven multitier reactor and fuel technologies, a single-tier fast reactor system, and a single-tier accelerator driven system. All of the approaches were evaluated with regard to concerns that the National Academy of Sciences brought up in its critical evaluation of transmutation technologies in 1996. Argonne concluded that transmutation can:

- Improve long-term public safety by reducing the radiotoxicity of nuclear waste below that of natural [uranium] ore within a period of 1,000 years as well as reducing the dose to future inhabitants by 99 percent.
- Provide benefits to the geologic repository by reducing the mass of commercial spent fuel by 95 percent and heat loads in the repository by 90 percent.
- Reduce the [plutonium] proliferation risk in commercial spent fuel by reducing the inventory of [plutonium] by 99 percent.
- Improve prospects for nuclear power by providing a viable and economically feasible waste management option for commercial spent fuel.[22]

Support

I have been a stalwart champion of nuclear fuel cycle research and development initiatives through the appropriations process. For example, Congress directed the DOE through the fiscal year 1999 Energy and Water Appropriations Act to study the accelerator transmutation of waste (ATW). First directed by the DOE Office of Civilian Radioactive Waste, Congress later turned the program over to the Office of Nuclear Energy, Science, and Technology. By late 1999, a joint accelerator program was created to combine the development of the accelerator production of tritium (for the nuclear weapons program) with the ATW program. This joint effort was named the Advanced Accelerator Applications (AAA) program. It was Congress's belief that the DOE should be focusing on a broader array of disposal options such as reprocessing, transmutation, and dry cask storage.

Interest in advanced fuel cycle technologies was revitalized by the George W. Bush administration. Under the courageous leadership of the Bush–Cheney team, a comprehensive energy policy was announced in May 2001 that promoted the study of the concepts I had been suggesting. The president's report stated: "[I]n the context of developing advanced nuclear fuel cycles and next generation technologies for nuclear energy, the United States should reexamine its policies to allow for research, development and deployment of fuel conditioning methods (such as pyroprocessing) that reduce waste streams and enhance proliferation resistance."[23]

In fiscal 2002, the DOE's program was expanded to research advanced spent fuel treatment technologies, as well as reactor and accelerator-based transmutation systems. The AAA program led the way in research on the UREX separation technology, discussed below. Regarding the transmutation program, research indicated that accelerator systems might be too expensive to build and operate. It further indicated that DOE should also investigate fast reactors, perhaps in concert with relatively small accelerator systems to help in the "final burn" of the long-lived radioactive elements.[24] The AAA program has since morphed into the AFCI, with greater emphasis on fuels and separations technologies. The AFCI program complements the department's international cooperative research venture regarding the future generation of nuclear energy systems, known as Generation IV (members include Argentina, Brazil, Canada, the European Union, France, Japan, the Republic of Korea, the Republic of South Africa, Switzerland, the United Kingdom, and the United States).

In 2003 the Senate Energy and Water Development Appropriations Committee designated over $77.8 million for the preconceptual design of an advanced recycle facility for performing research on scalable recycle technologies that are proliferation resistant, economical, and minimize environmental impact. This program was intended to study both reactor- and accelerator-based transmutation systems. The final agreement between the House and the Senate provided just over $58 million for spent fuel processing and advanced fuel cycles.

The program is moving forward with support from Congress and the Bush administration. For fiscal 2004, President Bush requested $64 million for the AFCI, the first time in a long time that a president requested funding for R&D on processing and recycling. As finally agreed to by both houses of Congress, the AFCI program will receive $68 million in funding, including money for the Idaho Accelerator Center. The DOE's programs have been refined to perform R&D on technologies that would affect the need for another repository in the United States. The goals are to:

- [e]nhance the design and reduce the long-term cost of the U.S.'s first geologic repository,
- [r]ecover the energy value of commercial spent nuclear fuel,
- [r]educe or eliminate the technical need for an additional repository,
- [p]ermanently destroy the plutonium that is contained within spent nuclear fuel, and
- [r]educe by a factor of 1,000 the radiotoxicity hazard posed by spent nuclear fuel.[25]

The government is breaking the work into two segments. In the near to intermediate term, that is, to 2015, work under the AFCI Series I will focus on developing proliferation-resistant separations technology to decrease the waste volume for the Yucca Mountain repository as well as developing fuels to enable the destruction of plutonium in light water or gas-cooled reactors by the middle of the next decade. It is hoped the research will lead to the conclusion that a second repository is not technically needed at the decision-making point in time. The next stage, AFCI Series II, focusing on R&D up to 2030, will develop advanced separation and transmutation technologies to enable the destruction of the long-lived actinides by fast neutron systems. This work is closely coordinated with the Generation IV Nuclear Energy Systems Initiative, which is studying candidate reactor systems for transmutation. The Los Alamos National Laboratory is taking the lead on developing transmutation science and technology and is developing a conceptual design for a transmutation target test facility.

In Dr. Richter's analysis, it may very well be feasible in the next five to ten years for the DOE to prove the viability of such transmutation concepts.[26] If the technology looks promising, then the government could then decide whether to proceed to a demonstration plant. The technology deserves the chance to be evaluated.

Conclusion

"Solving the waste management problem" is not only an ethical and noble goal, but an essential one if we are to enjoy the benefits of the expansion of nuclear power. But just what is meant by "solving" the waste issue? We have the technology, expertise, funding, and legislation to build the nation's first high-level waste repository. The WIPP facility for defense transuranic wastes has been operating for four years to date. Nuclear waste disposal is a political problem to solve, not a technical problem.

Opening the first geologic repository would demonstrate that waste disposal is not a barrier to the expansion of the nuclear industry. In a respected analysis of the necessary conditions for industry to make a commitment for a new nuclear plant, the Scully Capital consulting group concluded:

> In the unanimous view of utility and financial executives, no new plant will be undertaken without a permanent repository for spent fuel. Executives see the vote by Congress to proceed on Yucca Mountain as a bellwether indicator on nuclear power. More narrowly, the vote increases the likelihood that disposal risks will ultimately be resolved after NRC completes licensing and the facility is constructed.[27]

The DOE has studied the potential of various proposed technologies for spent nuclear fuel treatment and transmutation. The current once-through fuel cycle produces spent nuclear fuel that will take 300,000 years to decay to the toxicity of natural uranium ore from whence it came. As the DOE's 2003 status report on the AFCI concluded, "If the research suggested by the report should prove successful, application of advanced fuel cycle technologies would produce waste forms that would decay to that level after only about 1,000 years."[28]

Separation and transmutation technologies, as part of an integrated strategy for spent nuclear fuel management, could dramatically alter the radiotoxicity of final waste products destined for a repository and allow recovery of much of the residual energy in spent fuel. This option might involve systems utilizing both existing or new reactors, plus accelerators, resulting in a new fuel cycle. If this program is successful, we can recover the residual energy in spent fuel. We would also produce a final waste form that is no more toxic, after a few hundred years, than the original uranium ore. Plus it would make the fuel cycle less dangerous from a proliferation point of view. Research and development of these promising technologies is essential.

Nonproliferation benefits ensue from the fact that in the transmutation process, most of the plutonium is destroyed rather than stored in the repository as a tempting target for future would-be proliferators. Policymakers over the past decade convinced many that the once-through fuel cycle would be more proliferation resistant if spent nuclear fuel were to be buried. They argued that the intensely radioactive fuel would be a barrier to proliferation. Because this radioactivity comes mainly from fission fragments that will mostly decay in about one hundred years, this proliferation protection wanes relatively quickly. In reality, a repository full of spent nuclear fuel is a potential gold mine of weapons material. North Korea has already demonstrated that it can chemically separate the plutonium out of spent nuclear fuel. It would be far better to destroy this plutonium altogether before burying the nuclear waste.

Participation in international collaborative research programs on advanced fuel cycles and advanced reactor systems will allow the United States to regain the leadership position it lost when President Carter excused us from the table of fuel-cycle policy development. The advancement of our knowledge of safe and proliferation-resistant technologies that also improve our environment through safer waste disposal strategies is vital.

Critics of pursuing this closed fuel cycle argue reprocessing and recycling are more expensive than direct disposal as envisioned in the once-through fuel cycle. However, there may be significant savings in disposal costs, if, for example, reducing waste volumes and toxicity could save the $70 billion cost of another geologic repository. If the plutonium is reused in a reactor, then there would be tremendous resource savings attached to the closed fuel cycle.[29]

In conclusion, let me state that we must move forward with the licensing, construction, and operation of our nation's first commercial waste disposal facility. Public and Wall Street confidence demand nothing less. However, we must not hold hostage our nation's energy future to the plan to directly dispose of spent fuel. Now may not be the time to make plans for a reprocessing pilot plant or to put in place the transmutation scenario. However, now is the time to perform R&D on advanced fuel cycle treatments. While a consensus view on the fuel cycle has not yet developed, we must accelerate research to enable us to be in a timely position regarding the most beneficial technical approach to take in the nation's best interests.

10

The Case for Nuclear Power

The case for nuclear power is a compelling one, not only for its environmental benefits, but also for economic, nonproliferation, and energy security perspectives that support our national and international goals. Nuclear technologies significantly contribute to the health and safety of our nation's population through their application in the industrial, medical, and food supply arenas.

One: Energy Diversity Is a Moral and Human Development Imperative

The United Nations (UN) is projecting that the world's population will rise from its current level of 6 billion to 7.5 billion by 2020; the International Institute for Applied Systems Analysis (IIASA) is forecasting a population range of 8 to 12 billion by 2050. No matter what forecaster you choose to heed, it is generally conceded that the world's population will increase by 33 percent or more by 2050. There is clearly a case for the development of diverse sources of energy between now and then if the world is to survive.

We face a moral imperative to structure a diverse and environmentally sustainable system for providing energy for our nation and the rest of the world. Leaders from six of our country's finest national laboratories wrote to Energy Secretary Abraham in April 2003, eloquently stating this imperative:

> Energy is vital to human civilization. It underpins national security, economic prosperity, and global stability. As the world's most powerful and prosperous nation, the U.S. must lead the way in developing a diverse energy system that can

meet rapidly growing world energy demand in a way that promotes peace, prosperity, and environmental quality. This diverse energy system must include a growing component of nuclear energy.[1]

My perspective on sustainability is derived from my many years in the U.S. Senate, wherein I have devoted much of my time to environmental and energy concerns. When my Senate tenure started in 1973, the commemoration of Earth Day was three years young. During the ensuing years, I witnessed great strides toward the improvement of our nation's environment. We are uniquely fortunate to be prosperous enough to consciously choose to promote environmental concerns and conserve resources. However, we should focus on creating ways not only to continue these improvements in our own country, but also to assist other nations to improve their ability to protect the world's environment. The earth is the home we all share.

If it is accepted that people are the real wealth of nations, then human development is about creating an environment in which people can develop their full potential and lead productive lives in accord with their needs and interests. Human development is about expanding the choices people have to lead lives they value, and is about much more than economic growth.

While this may sound "theological," it is fact that we need energy to provide us with heat and electricity, to power our industries, our transport, and our modern way of life, and to maintain our standard of living. Energy is fundamental to meeting the human development challenges facing the world in the new millennium. Approximately 20 percent of the people in the developing countries, which make up about three-quarters of the world population, are illiterate; a billion people lack access to potable water sources; and 2.4 billion people lack access to basic sanitation. Nearly 325 million children do not receive an education at school and 30,000 children per day under the age of five die from preventable causes—that is, 11 million children annually.

According to the UN's International Labor Organization, 3 billion people—half of the world's population—live in poverty, with incomes of less than $2 per day. A quarter of the people in the developing countries earn less than $1 per day, and 60 percent live on less than $2 per day. Even in the Organization for Economic Cooperation and Development (OECD) countries, deprivation looms large, where more than 130 million are "income poor," 34 million are unemployed, and adult illiteracy averages 15 percent.

Figure 10.1 presents the relationship between the UN's "Human Development Index" and annual per capita electricity use. It shows clearly that the quality of human life is directly related to consumption of electricity. It can be seen, for example, that the developed countries have a human development index (HDI) of greater than 0.8, whereas undeveloped Africa ranges between 0.3 and 0.5. Scientist Alan Pasternak from the Lawrence Livermore National Laboratory found that the HDI reached a high plateau when a nation's people consumed about 4,000 kWh of electricity per capita.[2] It shows how far undeveloped countries must progress in their quest for a better quality of life for their people. Notice that in India, for example, with a huge and growing population, their people do not even consume half of the benchmark 4,000 kWh of electricity needed to reach an acceptable HDI.

One quarter of the world's population, about 1.6 billion people, have no access to electricity, and four out of five people who do not have electricity live in rural areas of developing countries, particularly in South Asia and sub-Saharan Africa. This lack of electricity exacerbates and perpetuates poverty in these countries. Lacking electricity, jobs cannot be created by industrialization of the economy.[3]

It is of interest to note that in 1999, with less than 5 percent of the world's population, the United States generated 30 percent of the world's GDP, consumed 25 percent of the world's energy, and emitted 25 percent of the world's carbon dioxide. The high electricity consumption rates in the United States, Canada, and Australia are, in addition to lifestyles, in part due to the large size of the three countries, and possibly the northern climates in the case of Canada and much of the United States.

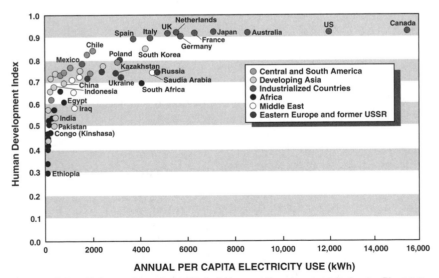

Figure 10.1.   Relationship of the Global Human Development Index to Electricity Consumption. *Source:* Alan D. Pasternak, "Global Energy Futures and Human Development: A Framework for Analysis," UCRL-ID-140773, U.S. Department of Energy, Lawrence Livermore National Laboratory, October 2000, 5.

By every human measure, the world "runs" on energy, and in the future the world will need more and more energy. Energy allows labor to be more and more productive and drives the machines that provide for the world's sustainable development. Sadly, in today's world, only one-quarter of the world's population has electricity, a resource that most people in developed countries take for granted. The alternative to an adequate and secure supply of energy is lack of development. History shows this results in people suffering the horrors of poverty, crime, disease, lack of education, and ultimately, premature death. An accompanying alternative to lack of development is chaos, violence, and anarchy that can lead to forced redistribution of wealth.

To meet unanticipated commercial and geopolitical challenges, America's energy should come from diverse sources. A diverse supply of energy is a strategic and economic imperative for the United States, and indeed, for all nations. We are only too well aware of the difficulties placed on the exercise of our foreign policy due to our dependence on imported oil. Our chemical industry is hamstrung as well. There are no substitutes for petroleum feedstocks, and the fertilizer industry is increasingly hurt by rising natural gas prices. The recent high level of natural gas prices is slowing the rate of economic recovery and job growth. Diversifying the sources of electricity generation, for example, with nuclear power generation can free up fossil fuels for uses that have no other substitutes, such as for petrochemicals, resulting in a cleaner environment, more U.S. jobs, and more efficient uses of raw materials.

As stated by Sir Bernard Ingham, if rational discourse prevailed, nuclear power would be advanced "as a major contribution to squaring the circle of man's sustainable development on this planet—providing sustainable power for its development without offence to its atmosphere and therefore further potential damage to its climate."[4] We must shift the dialogue from providing energy security to developing sustainable energy systems.

Two: Reduce Reliance on Oil and Natural Gas Imports

If the current trend for increased demand continues in many of the developed countries, the resulting increased dependence on energy imports could result in an increasing loss of control of costs and supply for the United States, as well as exposure to political instability. This could be true in other regions. For example, European dependence on fuel imports is in danger of rising from 50 percent in 1999 to about 70 percent in the next few decades, and in the case of oil and coal, could reach 90 percent and 100 percent, respectively.[5] Today, if China consumed oil at the same average rate as the rest of the world, it would consume more than the whole of Western Europe. If China's level of economic development rises to the level of that of its neighbor, South Korea, it will consume as much oil as Western Europe and the United States together. The global impacts could be very large if India, Africa, and Russia were to make similar advances.[6]

We must develop a balanced energy policy that includes a healthy reliance on clean nuclear power to counter the increasing competition for the supply of fossil fuels in the coming decades. China and India, with almost 40 percent of the world's population, rely primarily on domestic coal and domestic and imported oil. The IEA projects that China's petroleum imports could grow to 10 million barrels a day and India's to 5 million a day. Imports at this level would take over one-quarter of the world's projected increase in petroleum consumption.[7] By 2030, the combined net imports of oil into China and India are expected to more than double those into Japan, to be greater than those into Europe, and almost equal to those into the United States and Canada. Increases in China's natural gas imports will require a dramatic increase in gas supplies from the Middle East and Russia. Nearly half of the projected increase in natural gas demand in India will have to be imported as LNG.[8] Europe and possibly the United States will have to compete with these countries for supply, unless we turn to the nuclear option and diversify our electricity supply further.

America will experience an alarming reliance on fossil fuels during the next few decades and on imports of oil and petroleum products. The EIA projects total energy consumption to grow at about a 1.5 percent annual rate from almost 98 quads in 2002 to almost 136 quads by 2025.[9] Fossil fuel consumption is projected to increase from 84 quads in 2002 to 119 quads by 2025. To supply this demand, net crude oil imports rise from almost 20 quads in 2002 to 34 quads by 2025, an average annual increase of 2.4 percent.[10] If by 2025, as expected, the United States imports nearly one-quarter of all its natural gas from foreign sources, our country will head down the same dangerous road we started down fifty years ago with foreign oil imports.

The EIA has projected that total U.S. energy consumption will rise to 136 quads by 2025, with 53 quads coming from imports—that means 39 percent of our energy will be imported. In 2002, 29 percent of our energy came from imported sources.[11] While this might be accepted by a resource-poor country like Japan, it is unacceptable for the United States, which has vast coal resources and a well-developed nuclear technology capability. Nuclear power could contribute dramatically to providing nonfossil energy using fuel supplies available domestically. The time to act is now!

American consumers would reap economic benefits as well from reducing fossil fuel imports. At an energy subcommittee (of the Senate Energy and Natural Resources Committee) hearing on nuclear power generation, I heard testimony from the Energy Department's EIA regarding their recent energy study. It showed that generation of 6,000 MWe of new nuclear power would reduce both the consumption and the cost of natural gas by 3 percent in 2020. Such a 3 percent reduction in cost would save American businesses and consumers more than $3.6 billion annually.

A 1,000 MWe nuclear power plant saves the American consumer, each and every year, approximately:

- 14 million barrels of oil, domestic or imported, or
- 3.5 million short tons of coal, or
- 45 billion cubic feet of natural gas.

The exact savings would depend on the fuel quality, power plant types, and other related factors.

Three: Nuclear Technologies Improve Our Health and Safety

Nuclear technology not only provides us with electricity, it also saves lives every day. There are breast cancer survivors because radiation kills cancer cells lurking in the body after tumors are removed. In fact, there are more than 10 million diagnostic and therapeutic nuclear medicine procedures performed each year that save American lives.[12] And the safety of medicines that cure our diseases are tested by the Food and Drug Administration by using radioactive tracers.

I have consistently supported the funding for the DOE to provide a crucial cancer treatment provided by the radioactive material, bismuth-213, a rare emitter of high-energy alpha particles.[13] Its demand is expected to far outstrip our supply of this valuable cancer-fighting tool. One out of every three patients in hospitals receives either diagnostic or therapeutic treatment from nuclear medicines.

On yet another health front, food irradiation makes our supply of food safer and last longer. As I pointed out in my Harvard speech in October 1997, beef recalls, such as the one for 25 million pounds of beef from Hudson Foods due to contamination by E. coli bacteria, can be avoided if the beef is irradiated. Food irradiation can help cut into the alarming statistic that food-borne bacteria cause nine thousand deaths and one hundred thousand serious illnesses each year in the United States. There is no scientific danger stemming from food irradiation—only popular myth and scare tactics by consumer groups. Not only does irradiation kill life-threatening bacteria, it prolongs the shelf life of the food and thus we waste far less food every day.

Nuclear technologies help protect consumers in a variety of other ways. For example, you are safer when you fly in an airplane because of nuclear technology. The toughness of materials used in our airplanes is tested and assured through nondestructive nuclear testing procedures. It is also used to test the strength of products and components in the construction and manufacturing industries. It sterilizes consumer goods and medical supplies. The use of ionizing radiation protects us from bioterrorism; it was used to sterilize mail to U.S. senators after anthrax spores were sent through the mail to several congressional leaders in 2001. Indeed, nuclear technology is an essential aspect of our industrial infrastructure.

Nuclear technologies will help expand the frontiers of space, helping to lead us to untold discoveries about the beginnings of our universe. Current-generation technology allows the Mars *Spirit* and *Opportunity* rover robots to survive cold Martian nights and to analyze rock and soil data. It could play an even larger role in the future by allowing the next Mars rovers to remain on the planet for years instead of months by switching from the current solar power to radioisotope power sources.

Four: Nuclear Power Provides Electricity Safely and Securely

I could write a whole book on the subject of nuclear safety and all its aspects. However, I am only going to touch on nuclear's safety record, whether or not the threat of a potential criminal or terrorist attack should cause us to abandon nuclear power, and public perceptions of radiation and accident risks.

I believe the public's fear of radiation is at the base of many concerns about nuclear power. Some members of the public feel it is unfamiliar, unseen, and uncontrollable, and therefore something to be feared. Let me assure you, it is a natural phenomenon that has existed since the creation of our planet, of our universe. Rather than to be feared, it is a fact of life to be understood in context.

Every day of our lives, we are exposed to naturally occurring radiation from rocks in the earth, cosmic rays, radon gas, naturally radioactive foods, and even from our bodies. Some of us subject ourselves to radiation by smoking. Anyone traveling in an airplane receives doses of radiation due to flying at a very high altitude where cosmic rays are stronger. Ironically, most of us are quite willing to be subjected to man-made sources of radiation such as dental X-rays, medical procedures, and televisions. We accept radiation isotope production reactors because what they produce cures us of cancer or helps to diagnose an illness, yet we fear nuclear power reactors. I believe acceptance of one and rejection of the other is irrational. It is hard to have it both ways, especially since nuclear plants are so safe.

Returning to my thoughts on radiation, let's put it into perspective. The average person in the United States receives approximately 360 millirems of radiation, from all sources (natural and man-made), per year. For Americans, the greatest single source of radiation, 200 millirems per year, comes from radon gas. A round-trip cross-country flight gives the traveler 5 millirems of radiation. Most people accept the risk of receiving 30 millirems from a mammogram, 10 millirems from a chest X-ray, or even 1,400 millirems from a gastrointestinal series of tests. By comparison, living next to a nuclear power plant, one would receive less than 1 millirem per year. Figure 10.2 compares common sources of radiation.

While the public is willing to accept some radiation risks, as shown in the figure above, nuclear accidents and nuclear power plants are extremely low sources of radiation. However, the public has an extraordinary worry about the low-occurrence but high-risk scenarios. Even this concern needs perspective. In 1998, the Paul-Scherrer Institute in Switzerland analyzed actual energy industry-related accidents from 1969 to 1996. (This time period therefore included Three Mile Island and Chernobyl. There were thirty-one deaths from the Chernobyl accident and none from Three Mile Island.) The institute's database of 13,914 severe accidents included 4,290 energy-related incidents. The comprehensive data analysis found that for the energy sources of coal, oil, natural gas, liquefied petroleum gas, nuclear, and hydro, nuclear power is significantly safer than all other forms of energy generation. Natural gas, the next safest compared to nuclear, has a fatality rate ten times higher than nuclear. The report concludes that for each terawatt-year of energy use, the number of fatalities was 8 for nuclear, 85 for natural gas, 342 for coal, 418 for oil, 884 for hydro, and 3,280 of liquefied natural gas.[14] The significance of this comparison is that a terawatt-year is an extremely large quantity of energy, approximately equivalent to fifteen years of electricity generation in a country such as Canada.

Another comparison of nuclear power to other forms of energy demonstrates its comparative safety. In Richard Rhodes and Denis Beller's analysis, "equivalent lives lost per gigawatt generated (that is, loss of life expectancy from exposure to pollutants), coal kills 37 people annually; oil, 32; gas, 2; nuclear, 1."[15] Producing electricity from nuclear power appears to be the safest choice for our workers and the public.

The health and safety risks associated with nuclear power plants have been debated for decades. But when we debate this issue, we tend to ignore the risks we take in our everyday lives. Professor Bernard Cohen, from the University of Pittsburgh, compared the relative risks of nuclear power to other activities

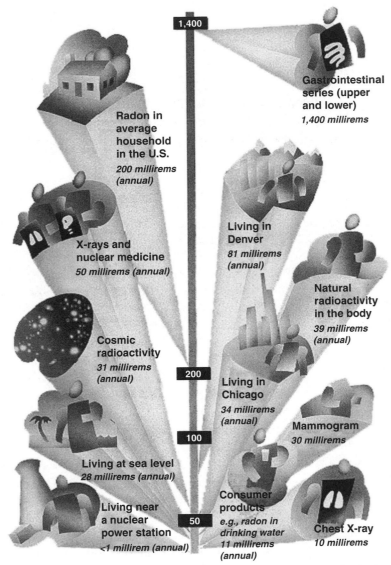

Gastrointestinal series (upper and lower)
1,400 millirems

Radon in average household in the U.S.
200 millirems (annual)

X-rays and nuclear medicine
50 millirems (annual)

Living in Denver
81 millirems (annual)

Natural radioactivity in the body
39 millirems (annual)

Cosmic radioactivity
31 millirems (annual)

Living in Chicago
34 millirems (annual)

Mammogram
30 millirems

Living at sea level
28 millirems (annual)

Living near a nuclear power station
<1 millirem (annual)

Consumer products
e.g., radon in drinking water
11 millirems (annual)

Chest X-ray
10 millirems

1,400

200

100

50

Figure 10.2.   Comparison of Radiation Emissions. *Source:* EPA, *Radiation: Risks and Realities,* EPA 402-K-92-004 (Washington, D.C.: Environmental Protection Agency, 2003).

using an increase of one chance in a million of premature death as the baseline for comparison. Here are some of the activities that he estimated would increase the risk of premature death by one in 1 million:

- Living for 5 years at the boundary of a nuclear power plant
- Living for 50 years within 5 miles of a nuclear power plant

- Traveling 10 miles by bicycle
- Traveling 300 miles by car
- Traveling 1,000 miles in an airplane
- Living 2 months in Denver, where there is naturally occurring radiation from granite and other rocks[16]

Professor David Ropeik of the Harvard Center for Risk Analysis confirms the point that the risks of radiation from a nuclear reactor are minimal as compared to risks faced by the U.S. general population each year. He reported that each year:

- Radiation from the sun causes 7,800 deaths from melanoma
- Medical mistakes kill as many as 98,000 Americans
- Food poisoning kills 5,000 Americans
- Air pollution kills 60,000 Americans[17]

Inevitably you will ask, "What about Three Mile Island?" Three Mile Island was the United States' worst nuclear accident, yet few realize that even though 90 percent of the fuel rods ruptured, the accident was a non-event from a radiation and health-hazard standpoint. The maximum exposure to the nearest member of the public was little more than a third of the NRC's annual limit for the public. And, no worker exceeded the commission's current annual limit for occupational exposure. In the Navy's nuclear submarines, the sailors who live and work within yards of operating reactors receive less whole-body radiation while underway than while at home and exposed to natural background radiation.[18] These examples demonstrate that there is some well-understood, quantifiable risk in operating a nuclear power plant. The data point to the fact that people accept far greater risks in their everyday lives without a second thought. These statistics, and the fifty-five-year operating record of the civilian nuclear power industry, show conclusively that nuclear power provides electricity in a safe and secure manner.

Turning now to the industry's safety record, it has over forty years of international operating experience. Its record in terms of safety-related events and radiation exposure to the public in the United States and worldwide continues to improve each year. The bottom line is that nuclear power is safer than fossil fuels from the perspective of accidents, environmental damage, and public health. New reactor designs being considered by the NRC make nuclear increasingly attractive, with standardized safety features and more inherent proliferation-resistance features. Even in terms of the wastes from all the energy options, the nuclear power industry is the only one required to dispose of all its wastes. The technology to reduce the volume and toxicity of such radioactive wastes is being developed and can further improve the safety of the nuclear option as compared to other energy options.

Finally, I would like to address a question often posed to me concerning nuclear power in the new environment of potential terrorist attacks on key parts of the modern world's energy infrastructure. The Electric Power Research Institute (EPRI) has examined the potential risks to public health and safety from both an air- and ground-based terrorist attack on a U.S. nuclear power plant. The aircraft crash impact study, using a Boeing 767-400 as the model, found that no parts of the engine, the fuselage, or wings, nor the jet fuel, entered the containment buildings. Neither the reactor containment nor the spent fuel storage pool structure were breached in the simulated attack. EPRI's ground-attack study concluded that the risks are indeed very low, well within the standards established by the NRC, and significantly less than risks encountered in everyday activities. For example, all causes of cancer in the United States result in 220 deaths per 100,000 people, accidents result in 35 deaths, and a hypothetical terrorist attack would potentially result in 0.004 deaths.[19]

The Center for Strategic and International Studies found in a 2002 national security exercise called "Silent Vector" that nuclear power plants would be a less attractive terrorist threat than other potential sites because of the physical security measures taken at nuclear power plant sites.[20] Not only has security and safety at plant sites been a priority in the past, measures mandated by the NRC and taken by the industry in the post-9/11 era have increased the safety of our nuclear plants.

While the risk is never going to be zero, I am totally confident in the safety and security measure in place protecting our nuclear power plants. We cannot afford to abandon nuclear power plants—just as we cannot afford to abandon skyscrapers or jet planes—because terrorists may threaten these symbols of our power and economy.

Five: Exemplary Nuclear Safety Record Mirrored by the U.S. Naval Nuclear Program

The U.S. nuclear-powered ships visit 150 ports worldwide in fifty foreign countries and dependencies each and every year. They are safe. They are welcomed. They are a symbol of just how safe nuclear reactors are.

The safety record of the U.S. naval reactor program is remarkable. In the fifty years that the Navy has been operating reactors, there has never been a reactor accident, nor has there been a release of radioactivity that has had an adverse impact on public health or the quality of the environment.[21] Over the years the Navy has operated 211 nuclear-powered ships and nine prototype nuclear propulsion plants. Today the Navy is responsible for operating 103 reactors (coincidentally the same number as commercially operating nuclear reactors in the United States) in 82 active nuclear-powered warships, 1 nuclear-powered deep-submergence vessel, and 4 training reactors. The Navy has amassed over 5,500 reactor years of operation without a reactor accident. By contrast, the commercial nuclear power plant operators have logged in 2,800 reactor years of operation.

With regard to its record of safe operations, it has been the Navy's policy to reduce personnel exposure to radiation from the nuclear propulsion plants and related facilities to as low as reasonably achievable. In carrying out this policy, the Navy has maintained personnel radiation exposure standards that match or are more stringent than those in the commercial nuclear power industry. The Navy frequently has taken a lead position in setting standards. For example, the current federal annual occupational radiation exposure limit of 5 millirems established in 1994 came twenty-seven years after the Navy's annual exposure limit of 5 millirems was established in 1967. This safety standard has been met. From 1968 to 1994, no civilian or military personnel in the Navy program exceeded its self-imposed 5-millirem annual limit. In fact, no personnel have exceeded 40 percent of the annual limit since 1980.

As I previously mentioned, crew operating the Navy's nuclear-powered ships receive less radiation exposure in a year than the average U.S. citizen does from natural background and medical radiation exposure. For example, the average sailor living onboard a nuclear-powered ship in the year 2002 was exposed to about 10 percent of the radiation received by the average American that year from natural background and medical sources. This standard is achieved because of the conservative shielding designs onboard these ships as well as the fact that crews are generally shielded from background radiation while at sea.

Safety is further ensured by the rigorous regulation, oversight, and inspection regime centrally controlled by an integrated headquarters organization of the DOE and the Navy. Self-audits and inspections are conducted at almost every level of the program. An important part of these reviews is evaluating the activity's ability to self-assess, in keeping with the principle that each activity must identify, diagnose, and resolve its own problems when auditors are not present to do so. This effort, along with other requirements, makes clear that day-to-day excellent performance must be the goal—and the norm— rather than merely "peaking" for an annual audit or inspection.

The strong regulatory oversight and inspection of the program by the Naval Reactors' headquarters' staff has long been held as a model for supervision of such a complex technology. In the early 1990s, the General Accounting Office

(GAO) performed an extensive and comprehensive fourteen-month investigation of environmental, health, and safety practices at the Naval Reactor programs' DOE facilities. At the conclusion of the review, the GAO testified before the DOE's Defense of Nuclear Facilities Panel of the House Committee on Armed Services that they had "reviewed the environmental, health, and safety practices at the Naval Reactors laboratories and sites and have found no significant deficiencies."

In addition to the GAO review, several knowledgeable organizations and oversight committees have recognized the program's excellent safety record. In 2003, the Columbia [Space Shuttle] Accident Investigation Board and the Defense Nuclear Facilities Safety Board both recognized the program as a model program with regard to technical and safety practices. In fact, the Columbia Accident Investigation Board cited the program as one of its major examples of "programs that have strived for accident-free performance and have, by and large, achieved it." The board went on to say that the program had a safety culture and organizational structure that made it highly adept in dealing with high risk by designing hardware and management systems that prevent seemingly inconsequential failures from leading to major accidents. The Navy program has always had the culture of working hard to solve and correct small problems before they grow into larger problems.

On the fiftieth anniversary of the Naval nuclear reactor program in 1998, Vice President Al Gore summed up its excellent safety record and culture:

> In the field of nuclear energy, not only has Naval Nuclear Propulsion made a contribution to national security of incalculable value, but has done so with a level of sustained excellence that is an outstanding example of Government serving its citizens. The Program's record of safety and environmental protection, started long before it was generally recognized how important these things are, is simply without equal.[22]

I am impressed with the Navy's ability to communicate its safety record to the local officials where its programs are located in the United States, and to the 150 ports in fifty foreign countries and dependencies where nuclear-powered ships call. Admiral Skip Bowman, fourth director of the Naval Nuclear Propulsion program, has done an outstanding job of building relationships with his regulatory counterparts at the state and local levels. The key to his success has been an insistence on a relationship that remains open to frequent communications, both to make the relationship stronger and to talk when small problems crop up. He has expanded his program's contacts and discussions to help make the local officials better understand the program. These counterparts are actively engaged in the Navy's radiological emergency planning business. This two-way open dialogue is exemplary and offers a message to all in the nuclear industry: acceptance of nuclear operations in communities around the country is positive when local officials see close-up just how safe this complicated technology can be.

Six: Nuclear Power Is the Only Noncarbon Baseload Source Available for Expansion Now

Baseload power supplies are the bedrock of a thriving economy. Many of our baseload electric plants in the United States are reaching retirement age. Just to maintain current supply capacity, new baseload plants must be brought on-line. To meet demand increases, more baseload power will be required. Which energy technologies are best suited for the job?

A crucial point is made by Richard Rhodes and Denis Beller about how we must deal with the here and now, not wishful-thinking options. As they pointed out, "because major, complex technologies require more than half a century to diffuse into global society,"[23] we must maintain and expand the options available to us today. Dr. Cesare Marchetti and Dr. Nebojsa Nakicenovic have extensively analyzed trends in primary energy substitution patterns. Figure 10.3 depicts this phenomenon. Basically, the message is that the replacement of primary energy technologies is a long process, lasting on the order of fifty to seventy years at the global level. The process leads to even higher energy efficiencies (lower energy intensities of economic activities) and lower carbon content (decarbonization). The dynamics of the process indicate that the next dominant energy sources will be natural gas, by midcentury, followed first by nuclear power and then by new renewables and/or fusion after the end of the century.[24]

Regardless of one's view of the future, and which energy technology would produce the "best" future, prudence demands that we make decisions today to keep our options open, and more importantly, focus on the sources of baseload energy available now—nuclear power, natural gas, and clean coal. Moreover, we must focus on the only source of electric energy that can provide baseload supply without any greenhouse gas emissions or supply constraints—nuclear power.

I am a strong supporter of solar and renewable energy. But, unfortunately, today they only represent a niche market—they are not capable of providing baseload power to our cities, hospitals, and factories. For renewables such as wind power to enjoy wide use in the United States, we must also have baseload nuclear and clean coal plants as well as gas-peaking plants. Nuclear and renewables are not mutually exclusive—in fact they must go hand in hand to

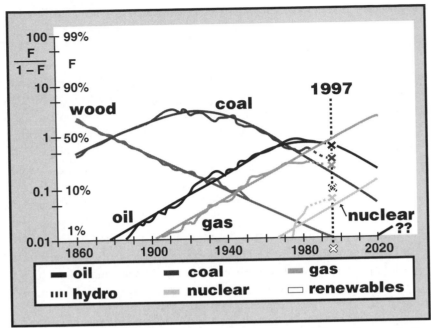

Figure 10.3. World Primary Energy Substitution. *Source:* Private Communication from Dr. Nebojsa Nakicenovic, December 14, 2003.

create a diverse and reliable electricity generation system. As we look at the situation in the developing countries, we see that as their populations surge and become more concentrated in urban areas, decentralized sources of energy will not be sufficient. Large baseload supplies of energy will be essential. If we want this to be clean energy, it should come from nuclear and clean coal plants.

Nuclear power plants, unlike renewables, are suited to provide baseload power for today's high-tech economies that require available power "24/7". Today's machinery and ultra high-speed communications processes demand high-quality power sources insulated from the vagaries of weather interruptions. The rapid urbanization of developing countries demands dense energy sources as well. Until reliable, low-cost energy storage options can be added to renewable sources, they are simply not suited to today's baseload energy needs.

Meeting future energy demand means improving energy efficiency and increasing our use of renewables. But the reality is that the renewable sources of energy have attendant environmental and efficiency issues. Expansion of some renewables is constrained. Developing countries' use of biomass is rapidly depleting the world's forests. Hydroelectric power is almost fully utilized. Ocean tides and thermal ocean gradients have limited potential. Geothermal energy, solar thermal, and solar photoelectric will contribute, as will wind power. But all of these together can probably not account for more than 25 percent of our total energy requirements, if that.[25]

Renewable energy sources may be seen as "green" and "more acceptable" to some in the public, particularly among those who consider themselves to be environmentalists, but when developers attempt to introduce them on a large scale, people resist having them in their backyards. Take for example the hullabaloo over Cape Wind Associates' attempt to build a massive farm of 130 windmills just seven miles off the coast of Hyannis in Cape Cod Sound. You would think it would be compelling that such a wind farm could provide Cape Cod residents with up to 75 percent of their electricity without emitting a single drop of carbon dioxide, nitrous oxide, or mercury, without consuming a single drop of tainted Middle Eastern oil! But no, the residents are up in arms over the potential damage to birds that could be sucked into the wind turbines, or the threat to residents' pristine view on a beautiful summer day, or the problems sailors could encounter from changing wind patterns. "Green wind power" may not be the savior many hoped it would be.

Renewables are in the early days of their development and thus the public is not yet aware of their relative drawbacks. Renewables are in their heyday, but will be increasingly subjected to the same antigrowth, antitechnology environmental pressures that more fully developed power sources such as nuclear are.

Relying only on conservation of energy, improved efficiency standards, or huge increases in power supplies from renewable sources of energy, rather than increasing the use of green nuclear power, would limit the economy's flexibility. This would be especially inappropriate at this time, as we have insufficient information about the practical potential of the other options such as renewables, clean coal, or fusion, and we have certain knowledge about the future demand for energy. The only rational approach to providing clean, economic supplies of energy is to keep open as many options as possible. We must rely on the nuclear option until far more is known about the other energy options.[26]

Seven: Nuclear Power Is an Increasingly Competitive Option

Nuclear power units are presently in direct competition with other forms of electricity generation, primarily coal-fired baseload units. Operating and maintenance costs at nuclear plants are at their lowest levels in more than ten years. Average availability factors are at an all-time high and safety records are excellent. Average nuclear and coal-fired operating costs in the United States continue to be nearly identical, with nuclear perhaps having a slight edge. As depicted in figure 10.4, the EPRI has shown that during the year 2000, these costs remained just under 2.0 cents/kWh, with average production costs of 1.76 cents/kWh reported for nuclear units, 1.79 cents/kWh for coal-fired plants, 5.69 cents/kWh for natural gas, and 5.28 cents/kWh reported for oil-fired plants.[27] The Nuclear Energy Institute's (NEI) comparison of production costs shows nuclear at 1.71 cents, coal at 1.85 cents, gas at 4.06 cents, and oil at 4.41 cents.[28]

Not only is nuclear cost competitive with coal and gas, it will have an important moderating effect on both the percentage and the cost of imported fuels, particularly natural gas. At hearings before congressional committees in 2003, Federal Reserve Chairman Alan Greenspan testified that increased domestic production, diversification of U.S. energy supplies, and short-term imports of liquid natural gas are the only measures that will ease the tight supply and rising prices of natural gas. To expand its energy base, Chairman Greenspan noted that the United States should start looking at nuclear power to do so.[29] Chairman Greenspan said, "I've always testified in favor of re-examining what I think is a policy which is mistaken, namely our views toward nuclear power. I do think that the technologies have improved immensely. . . . I do think an overall policy of energy cannot dismiss the issue of

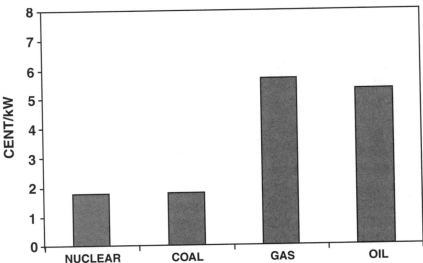

Figure 10.4.   Comparative Energy Costs. *Source:* T. U. Marston, "Nuclear Power's Role in Meeting Environmental Requirements: Evaluation of E-EPIC Economic Analysis," Electric Power Research Institute, final report, Palo Alto, California, January 2003.

nuclear power."[30] Chairman Greenspan recognizes that "the advantages of nuclear power are very considerable."[31]

A comparison of seven generating cost estimates prepared in Europe during the period 1997–2003 for nuclear power and gas turbines shows costs of 2.4–3.9 euro cents/kWh for nuclear and 3.1–9.2 euro cents/kWh for gas turbines. Each of these studies demonstrated the competitive cost of nuclear power in Europe compared to fossil fuel technologies.[32] According to a recent study by the Federation of Electric Power Companies of Japan, for electric power plants beginning operation in 2002, with a minimum forty-year lifespan and an assumed 80 percent capacity, the projected generating cost for nuclear is 5.6 yen/kWh, 5.9 yen for coal, 6.3 yen for liquefied natural gas, and 10.95 yen for oil.[33]

While cost comparisons can be tricky, and sometimes skewed by their underlying assumptions, clearly in the United States the operating cost of nuclear power includes the government's one-tenth of 1 cent/kWh waste disposal fee, while the corresponding fact is not the case for coal or natural gas. The price that consumers pay for their nuclear powered electricity internalizes *all* of its costs—spent nuclear fuel disposal, radioactive waste management, decommissioning, and safety regulation. Costs for other sources do not typically include these counterpart components.

A study by the Royal Engineering Academy of the UK has shown that nuclear power is the lowest cost alternative for generating electricity, especially when the cost of taking care of carbon dioxide emissions is included. The study estimates that the production costs in the UK for nuclear plants is 2.3 pence/kWh (based on the assumption of 1,150 British pounds/kW construction cost). Emission of carbon dioxide by coal-fired plants adds a premium of 2.5 pence/kWh, so that the cost of electricity generation by coal would nearly match those of onshore wind generation at 5.4 pence/kWh.[34]

Nobel laureate and professor Burton Richter compared U.S. nuclear power capital, operating, and waste sequestration costs with those for fossil fuels. In order to be fair to each technology, Dr. Richter's comparison took into account the fact that spent nuclear fuel and fossil fuel need waste sequestration. SO_x, NO_x (sulfur and nitrogen oxides), and greenhouse gases must all be considered waste products that must be dealt with. His study concluded:

Estimates on the cost of sequestering the carbon from fossil-fueled power plants are more uncertain, because the R&D is in an earlier phase. . . . Cost estimates for sequestration are 1 to 1.5 cents per kilowatt hour for gas-fueled power production and 2 to 3 cents per kilowatt hour for coal-fueled power production. . . . Including a sequestration charge on fossil fuels gives a quite different picture of the relative costs of different sources of electricity. Nuclear goes from being the high-cost source to being comparable to natural gas and cheaper than coal.[35]

Table 10.1 presents Dr. Richter's comparison.

Table 10.1.   Power Costs (U.S. Cents per kWe-hour)

Category	Nuclear	Coal	Gas
Capital and operating cost	4.1–6.6	4.2	3.8–5.6
Waste sequestration	0.1	2.0–3.0	1.0–1.5
Total	4.2–6.7	6.2–7.2	4.8–7.1

Source: Burton Richter, "A Personal Commentary on the MIT Nuclear Power Study," private communication, March 15, 2004.

The million-dollar question is, of course, whether advanced nuclear power plants will be competitive with coal and natural gas plants that have shorter construction lead times. A lot depends on the extent to which U.S. government policy can help with the start-up costs of the first advanced generation plants or whether the Energy Policy Act's provisions I championed regarding production tax credits for nuclear (and renewables, by the way) will be passed by the U.S. Congress.

Another way to level the playing field for new nuclear power plants is through the imposition of carbon taxes on coal and gas plants. MIT's analysis shows that if the government enacted a policy of reducing carbon emissions through a tax, nuclear power becomes an economically competitive baseload option. The MIT study reaches the conclusion that even at the far end of its range for nuclear power costs (6.7 cents/kWh), nuclear starts to be economic when the carbon tax on coal or gas reaches the level of $100 per ton of carbon. If we assume the costs of new nuclear power plants will be less than the high forecast because of improvements in the licensing and construction processes, then nuclear would be the preferable option at a tax level of $50/tC.[36]

To remain competitive in the future, the nuclear industry must address potentially high initial capital costs and long construction lead times so that new nuclear power plants can be as competitive as those already online. Standardization of plant designs must be carried out as planned by the industry. Some of these high costs are inevitable with the introduction of advanced nuclear systems that the NRC has only just begun to license. The nuclear industry and government need to develop an acceptable formula whereby these first-of-a-kind costs can be shared so that they do not unfairly jeopardize new advanced nuclear plant orders. Shortened and thus less-costly construction schedules due to improved construction techniques such as modular construction, as well as a streamlined one-stop licensing process now in play at the NRC, should eliminate the severe delays witnessed in the past that served to drive up nuclear plant price tags.[37]

Eight: Nuclear Power Is Compatible with the Hydrogen Economy

Twenty-first century energy production systems will need to produce more than just electricity. The economies of tomorrow will require hydrogen, fresh water, and process heat in quantities that can meet centralized or distributed demand in an environmentally acceptable manner. The benefit of using hydrogen as a transportation fuel is significant, as over one-third of carbon dioxide emissions in the United States come from the transportation sector.[38] Switching transportation fuels will have a huge environmental benefit for our country, especially if produced by a green form of energy.

While any source of electricity can be used for electrolysis of water and hydrogen (H_2) production, the trick will be to do it cleanly and economically. Furthermore, conventional electrolysis is limited in efficiency.

Since hydrogen will unquestionably be the transportation fuel of choice for the long term, barring some revolutionary new technology (such as clean, low-cost coal gasification), hydrogen should be produced by electricity from nuclear power. It is the only option, barring a scientific breakthrough that can possibly meet the scale of deployment required at a low cost while meeting the goals of providing energy security and protecting the global environment. Nuclear power systems meet the criteria demanded for hydrogen production, providing high temperatures at low cost as well as a reliable source of energy. Nuclear power plants typically operate for one to two years without stopping for refueling. Their large scale is also compatible with the large power needs for hydrogen production. It makes no sense whatsoever to produce hydrogen for environmentally friendly hydrogen-powered fuel cells using environmentally unfriendly fossil fuels.

A 700 MWe nuclear power plant can produce sufficient hydrogen from electrolysis to power about 650,000 cars.[39] Paul Grant, a science fellow at the EPRI, has estimated that the United States would need about 84 MMT per year (or 230,000 MT of hydrogen daily) to fuel all of our automobiles. It would require about 400 GWe of electric capacity to extract this much hydrogen.[40] However, this would eliminate the consumption of 9 million barrels of oil per day, which could be a huge boon to our energy and national security. If we were to produce this much electricity using fossil fuels, it would require hundreds of coal or natural-gas-fired power plants, which would spew tons of greenhouse gasses into the atmosphere.

Clearly the trade-off would provide large cost savings and environmental benefits.

The DOE, under its nuclear hydrogen initiative program, will conduct research into advanced thermochemical technologies. The DOE's ambitious schedule calls for lab testing of two nuclear candidate processes in 2005, initiation of construction of two pilot production plants in 2006, operation of pilot plants starting in 2008, and followed by the construction of engineering-scale hydrogen production systems by 2012.[41] These technologies hold out the promise that, when used in tandem with next-generation nuclear energy systems under development now, they would enable the United States to generate hydrogen at a scale and cost that would support the hydrogen economy in this century.

To this end, I proposed the construction of an advanced nuclear reactor at the Idaho National Engineering and Environmental Laboratory (INEEL) to demonstrate both electricity and hydrogen production. The Senate subcommittee that I chair had recommended an appropriation of $8 million to support the necessary R&D for the development of the high-temperature electrolysis and sulfur-iodine thermochemical technologies necessary to the advanced reactor hydrogen cogeneration project at the new Idaho National Laboratory (INL) which will subsume parts of INEEL in the next year.[42] While all the funding I sought was not provided for this program, the fiscal 2004 energy and water appropriations legislation signed by the president earmarked $6.5 million to it. In addition, the legislation provided an initial $15 million to begin the design work on an advanced hydrogen cogeneration reactor. This new reactor will feature improved safety, reduced waste, higher efficiency, and increased proliferation resistance and physical security. It will produce hydrogen and generate electricity. The advanced reactor hydrogen cogeneration research project will move America toward an advanced nuclear energy and hydrogen economy. The project will put the nation back on track to develop better reactors while also responding to the president's call to develop new, environmentally friendly ways to produce hydrogen for a hydrogen economy.

The concept of a hydrogen economy is relatively new to the American consumer, although the United States already uses 11 tons of hydrogen each year in the fertilizer, chemical, and oil industries. Almost all of this hydrogen is produced by the steam reformation of natural gas. Unfortunately, for every kilogram of hydrogen currently produced by this method, 10 kilograms of greenhouse gasses are released into the atmosphere. Rather than methane and natural gas, water would provide an inexhaustible source of hydrogen if the oxygen in it could be separated out through the thermochemical water splitting process or high-temperature electrolysis. These processes would require the application of high-temperature heat—as high as 800°C to 1,000°C. The clincher is that huge amounts of energy are needed to separate the hydrogen and oxygen.

Unfortunately, the current generation of LWR nuclear power plants produces coolant outlet temperatures of approximately 400°C, too low for use in thermochemical cycles for either direct splitting of water or steam methane reforming.[43] While the sodium-cooled fast breeder reactor peak coolant temperature of 600°C would couple with steam reforming or electrolysis, it may not be suitable for water cracking. However, gas graphite reactors have peak coolant temperatures in the range of about 900°C and are suited to water splitting.

For example, the South African pebble bed modular reactor (PBMR) has a coolant discharge temperature of about 950°C; and General Atomic's proposed very high temperature reactor is also expected to be suited to water splitting. Other countries are also considering development of high temperature reactors for thermochemical water cracking. The Japanese 30 MWe high temperature reactor is operating reliably today. At this early stage of development, it looks like the thermochemical water splitting process using a sulfur-iodine cycle for producing hydrogen is the most promising and this process is now operating, on a bench scale, in Japan. While the choice of technology may not be certain at this time, the bottom line is that new reactor technologies are needed to support the hydrogen economy.

What is demonstrably clear is the need for government support in the early stages of our crash program to replace gasoline with hydrogen as the world's primary transportation fuel. The conventional thought that nothing will ever dislodge oil/gas as our primary fuel must be turned on its head. Thankfully, in February 2003, President Bush announced a hydrogen fuel initiative with a multiyear $1.2 billion budget to develop hydrogen production technology.

Secretary of Energy Spencer Abraham called for an international effort to produce hydrogen with energy provided from a whole host of energy sources, including nuclear, natural gas, clean coal, and renewables. For the United States, the Bush administration's goal is to create a mass-market penetration of hydrogen fuel cell cars by the year 2020. The interim success of hybrid gas-electric automobiles that are currently beginning to appear on U.S. roads could result in slowing up deployment of hydrogen-fueled vehicles.

Progress is evident. There will be a competitive public solicitation to receive proposals to build a hydrogen cogeneration reactor at the INL. Framatome reportedly has plans to respond with a proposal to develop a commercial very high temperature reactor in the modular high temperature gas-cooled reactor family. Westinghouse/BNFL plans to compete with an adaptation of the PBMR design, and General Atomics is working on a variation of its gas turbine modular helium reactor for the INL project.[44]

Mine is not a Jules Verne-vision. In the words of distinguished U.S. scientists Drs. Leon Walters, David Wade, and David Lewis:

> The transition to a nuclear/hydrogen economy can begin today with the current generation of nuclear reactors and the mature electrolysis technology. Nuclear energy can support a cleaner use of fossil fuel through nuclear assisted steam methane reforming for heavy oil enhancement. Finally, highly efficient carbon-free means of producing hydrogen by coupling advanced high temperature nuclear reactors with thermochemical processes is [sic] on the horizon.[45]

Nine: Nuclear Power Is a Green Energy Option

As pointed out by Richard Rhodes, Shippingport, the first nuclear reactor built in the United States, was built for Duquesne Power and Light because the residents of Pittsburgh were fed up with the legendary pollution afflicting their city.[46] They opposed construction of a coal-fired plant and the sulfur it would have added to an already too-smoky city residents were trying to clean up through what the city elders referred to as "Renaissance I." In 1954, Duquesne submitted the winning one of nine proposals to the Atomic Energy Commission (AEC). The deal was that Duquesne would contribute the turbines and an additional $5 million in funding for reactor plant and related structures, and agree to buy the steam heat from Shippingport at a price slightly higher than that from its other plants; the AEC would provide everything related to the nuclear power side of the plant.[47] While not labeled as such in that day, the country's first nuclear power plant was built because it provided green energy to a city choking on pollution from coal-fired plants and factories.

More than fifty years later, and after decades of resistance by "environmentalists," a consensus seems to be developing across industry and academic communities that the potential for environmental catastrophe can be avoided if nuclear power plays a vital role in providing clean energy. The opening paragraph of the prestigious MIT study concluded:

> We decided to study the future of nuclear power because we believe this technology, despite the challenges it faces, is an important option for the United States and the world to meet future energy needs without emitting carbon dioxide (CO_2) and other atmospheric pollutants. Other options include increased efficiency, renewables, and sequestration. We believe that all options should be preserved as nations develop strategies that provide energy while meeting important environmental challenges.[48]

Nuclear energy has advantages in terms of global warming, cost stability, and high capacity factors that make it compatible with the goals of sustainable development for tomorrow's world. The World Energy Council's research and global studies lead us to believe that nuclear energy will play an essential role in electricity production and in strategies to combat global warming. For base-load electricity generation, one of the most effective means currently available to reduce CO_2 emissions is nuclear power. Countries with the highest proportion of nuclear and/or hydropower have the lowest CO_2 emissions per kWh.[49]

Nuclear energy has an advantage over other carbon-free energy technologies in that it possesses a very high energy density sufficient to produce very large amounts of electricity or hydrogen in a facility with a small footprint (small volumes of discharged waste).

Nuclear energy has a very low life-cycle carbon emission when compared to other energy sources. As seen in figure 10.5, a comparison of CO_2 emissions from various power sources in Japan shows that, in kg of CO_2 emissions per kilowatt-hour, hydroelectric is lowest at 0.02, with nuclear right behind it at 0.022. For comparison, coal-fired electricity produced the highest emission of carbon at 0.99, while photovoltaic solar and wind emitted 0.06 and 0.04, respectively.[50]

Coal combustion emits almost twice as much carbon dioxide per unit of energy as does the combustion of natural gas, whereas the amount from crude oil combustion falls between coal and natural gas emissions. The combustion of

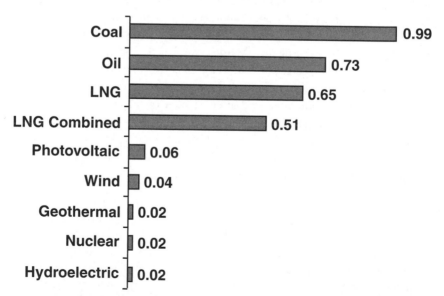

Figure 10.5. Comparison of Carbon Dioxide Emissions Intensity by Electrical Generating Fuel in Japan (kg CO_2/kWh). Total CO_2 emissions from mining, fuel transportation and refining, plant construction, operations, and maintenance. *Source:* From the report of the Central Research Institute of Electric Power Industry, "Life Cycle Analysis of Power Generation System," March 1995.

1 pound of coal, with a 78 percent carbon content and a heating value of 14,000 BTU per pound, will generate 2.86 pounds of carbon dioxide.[51] Emissions rates of CO_2 and other gasses for coal, oil, and gas are given in table 10.2.

Table 10.2. Fossil Fuel Emission Rates (pounds per megawatt-hour)

	Coal	Oil	Gas
CO_2	2,186	1,618	1,186
SO_2	10.85	8.22	0.25
NO_x	4.79	3.72	1.72

Source: Data from eGRID 2002PC, Version 2.01, Environmental Protection Agency (2000 data used).

Nuclear energy provides the United States with its greatest share of clean, emissions-free energy today, almost three-quarters of the United States' clean electricity supply. As nuclear continues to expand, each new nuclear power plant avoids about 8 million tons of greenhouse gas emissions—CO_2, SO_2, and NO_x—each year. Consider nuclear energy's track record:

- For carbon mitigation alone, the 103 nuclear power plants in the United States today, alone, displace 174 million metric tons carbon (MMTC) per year (1 ton of carbon is equivalent to 3.67 tons of CO_2).
- Since the start of commercial nuclear-powered electricity in the 1960s, over 10 billion tons of greenhouse gas emissions have been avoided.[52]
- At the close of the last century, nuclear power plants provided about half of the total carbon reductions achieved by U.S. industry under the federal voluntary reporting program.
- Between 1973 and 2000, U.S. nuclear plants reduced cumulative emissions of nitrogen oxide and sulfur oxide by 33.6 million tons and 66.1 million tons, respectively. Cumulative carbon emission reductions were 2.79 billion tons.
- Worldwide, the fleet of approximately 438 nuclear power plants reduced carbon emissions by 500 MMTC.
- Because of nuclear power, the world today avoids the emission of approximately 1,800 million tons of CO_2, 15 million tons of sulfur dioxide, and 7 million tons of nitrogen oxide, each and every year.[53]

Nuclear energy is essentially emission-free. The inescapable fact is that nuclear energy is making an immense contribution to the environmental health of our world.

Consider the future: as the population increases, energy consumption per capita increases, and the overall increase in energy consumption is thus magnified. It's a recipe for disaster if nuclear power does not increase its role in providing clean energy for the world. The European Commission (EC) recognizes the rising energy demand situation and its consequences. It estimates that world energy demand will increase at about 1.8 percent per year between 2000 and 2030. It projects that CO_2 emissions will increase by 2.1 percent per year on average, and that world CO_2 emissions will reach 44,000 million metric tons annually by 2030. The EC has projected world oil production will increase by about 65 percent, to 120 million barrels per day by 2030.[54]

Under this growth scenario, nuclear power must be deployed in greater quantities to reduce the otherwise inevitable greenhouse gas emissions due to increased energy supply. Richard Rhodes puts this into perspective:

> In a world where everyone is supplied with half the energy Americans take for granted today, by 2050 let's say, primary energy supply would grow to 1,000 quads per year compared to 283 quads in 1995. Simply to stabilize the atmospheric carbon dioxide concentration by then, the fossil fuel component of that 1,000 quads will have to drop by 22 percent compared to 1995. Carbon-free components—hydropower, nuclear, renewables, sequestration—will have to increase by more than 1,300 percent.[55]

President Bush has proposed an 18 percent voluntary reduction in domestic greenhouse gas intensity of our GDP by the year 2012. To help meet this goal, the U.S. nuclear industry has committed to adding 10,000 MWe of expanded capacity. The emissions avoided by this incremental increase alone—an estimated 22 MMTC equivalent—represent approximately one-fifth of the president's carbon-reduction goal.[56] I believe that U.S. industry plans are realistic. The goal can be met by a 3,000 to 5,000 MWe increase in productivity in current plants, power uprates through capital investments and steam generator replacements, and the restart of TVA's Browns Ferry reactor.

European Union Commissioner Loyola de Palacio recognized the role that nuclear power could play in helping Europe meet its national and international obligations made in Kyoto to reduce greenhouse gases. In a speech in 2002 she stated:

> Against the background of electoral campaigns, Europe needed a rational, objective and transparent debate because nuclear energy is a factor for stabilizing prices and guaranteeing supply. A new problem exists: *the speeding up of global warming*. Nuclear energy would allow us to cut down on 300 million tons of carbon every year in Europe. If we abandon nuclear energy, we have to say how we will produce this electricity and tons of CO_2 will be produced with conventional forms of energy. It's a very serious problem. Without nuclear power, Europe will be unable to meet the Kyoto objective.[57]

To summarize, when nuclear is compared to other forms of energy its environmental advantages shine. It has no greenhouse or other air-polluting emissions; it requires very little land; its wastes will be disposed of (and the disposal costs are collected from the producers, unlike all the other sources); and it has a very minor ecological or aesthetic impacts. Table 10.3 compares all the energy sources and points to nuclear's overwhelming environmental advantages.

Ten: Desalination and Potable Drinking Water Production

Nuclear energy can also help address one of the world's most pressing needs: potable drinking water. There is a golden opportunity to use nuclear power for water desalination. We do that today onboard nuclear submarines. With the advent of new desalination systems such as reverse osmosis, the potential is there.[58] The French CEA is studying a 300 MWe cogeneration nuclear plant for electricity and desalination. With the 300 MWe plant, 250 MWe are used for electricity production and 50 MWe for desalination, producing 200,000 cubic meters per day. This is enough water for 1 million people.[59] Similar desalination work is in progress in Korea, Kazakhstan, and Japan. China reportedly will construct a nuclear power plant in Yantai City to desalinate 35 million gallons of water per day. India is developing a desalination plant at its PWRs at the Madras nuclear power plants in Kalpakkam.[60] Millions of people globally could have access to clean, desalinated water because of nuclear technologies.

Conclusion: Imaging Life without Nuclear Power

To those of you who advocate turning back the clock on nuclear power, let me pose the key question about the future of nuclear power in the United States: What would replace nuclear power plants and at what economic and environmental cost?

U.S. energy demand will grow at an estimated 1.5 percent rate through 2025. U.S. electricity demand will grow by 53 percent over the next twenty years. Current-generation power plants will be retired in due course. If the EIA's estimates are correct, the United States will need a total of 335 GWe of new power plants. Fuel-efficient technologies and conservation will not bring down the growth in energy demand to the point that new plants are not needed. Fuel-efficient cars will not decrease the demand for electricity, but only for domestic oil and petroleum imports. In fact, moving to a hydrogen fuel economy for transportation will only serve to substantially increase the need for electricity.

Table 10.3. Comparison of Environmental Impacts of 1,000 MWe Power Plants

Alternative	Average U.S. Emissions (lbs/MWh)	Land Use (acres)	Waste	Ecology
Wind	None	150,000	Minor	Serious disruption of bird and bat habitation
Coal	CO_z—2,186.38 SO_2—10.85 NO_x—4.79	1,700	Mercury, radioactive fly ash, major amount of solid waste	Habitat loss, especially from acid rain; waste heat
Oil	CO_z—1,618.97 SO_2—8.22 NO_x—3.72	120	Scrubber sludge particulates	Habitat loss, especially from acid rain; waste heat
Natural gas	CO_z—1,186.58 SO_2—0.25 NO_x—1.72	110	Some radionuclides	Some habitat loss; waste heat
Hydro	None	1,000,000	Minor	Serious disruption of natural habitats, especially of fish; loss of agricultural land
Solar thermal	Minor emissions	14,000	Minor	Avian mortality
Photovoltaic cells	None	35,000	Toxic chemicals in cells: arsenic, cadmium, and gallium	Habitat loss
Nuclear	None	213	Radionuclides	Waste heat

Source: Based on generic environmental impact statement for license renewal of nuclear plants (NUREG-1437, Vol. 1), table 8.2, and eGrid2002PC, Version 2.01, Fossil Fuel Emission Rates, Environmental Protection Agency, 2000 data. Emissions data rounded up.

If we maintain the status quo in which fossil fuels provide about 65 percent of electricity generation, and we need significant numbers of new power plants,

- our imports of oil and natural gas would surge,
- greenhouse gas emissions would explode from burning ever more amounts of coal, and
- we would have to allow more domestic coal mining as well as oil drilling with its attendant environmental and political issues.

If we replaced 20 percent of the electricity generation provided by nuclear power today with wind farms, for example, imagine what the United States would look like with the number of windmills needed to produce this energy. The equivalent amount of installed wind capacity replacing nuclear's 20 percent share of the new capacity needed would be approximately 205 GWe.[61] Based on the American Wind Association's calculation that on a flat terrain 50 acres of land are required per MWe of installed capacity,[62] wind farms would occupy 10.25 million acres of land or comparable areas offshore—about 16,000 square miles.

To those who want to shut down the current fleet of nuclear plants and/or not build any more plants, I ask, what would replace these plants and what are the costs of these alternatives? If nuclear power plants were shut down today, we would need to take more than 135 million cars off the roads to maintain carbon emissions at current levels. The United States would have to import more oil and gas supplies, explore and drill on U.S. soil and coasts for oil, gas, and coal, suffer increased pollution from burning fossil fuels or expend considerable amounts of money for removing the pollutants, and pay higher bills for electricity produced from renewables. These higher energy bills or perhaps constrained power supplies would drag down our economy.

Alternatively, the expanded use of nuclear energy could protect the environment, provide the developing nations with improved control over their economies, and provide today's underdeveloped countries with the means to raise their standard of living toward that of the rest of the world. Consequently, nuclear power would contribute to promoting global peace in the world. It is up to all of us to ensure that nuclear power plays this role in the drama of human development because of its potential contributions, because if it cannot, the world will be a sorrier place in the future.

The outlook for nuclear power both here and abroad is positive. Ringing the death bell is premature to say the least. Robust signs of the nuclear option abound, including:

- growing awareness that we must provide a diverse energy supply to ensure national security, prosperity, and global stability;
- increased efficiency, extended nuclear plant life, and power uprates leading to an overall increase in electricity generation by nuclear power despite no new nuclear plants coming online in the past decade;
- competitiveness of the current generation of nuclear power plants with the reasonable expectation that the next generation of plants can and will be built at an affordable capital cost on a timely basis;
- development of new nuclear power plant designs with the promise of safer, more efficient, and increased proliferation-resistant;
- government R&D on advanced fuel cycles with an eye toward developing fuel cycles with less waste and more proliferation resistance;

- movement toward the licensing and construction of the first geologic high-level waste disposal facility and operation of the WIPP facility;
- signals that U.S. utilities are taking the first steps toward new nuclear plant orders;
- development of a national energy policy favorable to nuclear power;
- development of nuclear power systems for space travel to Mars and beyond;
- ever-widening use of radioactive technologies in industrial applications ranging from autos to aerospace; and
- the increasing use of nuclear medicine for safe and effective medical procedures.

These positive signs are the backdrop for a compelling case for the future role of nuclear power. Nuclear power is available today and can meet the increasing demand for electricity without having to rely on uncertain technology. It plays a vital, diverse role in assuring the health and general welfare of the American people through medical and industrial applications. Nuclear energy is the only baseload energy source that can meet increasing energy demand while reducing harmful greenhouse gas emissions in the process. The nuclear option is competitive now with other forms of energy that do not yet internalize all of their environmental and social costs, providing an affordable source of energy to drive our economic growth. It will play a key role in reducing our reliance on imported fuels, especially when it helps usher in the hydrogen economy in a way no other source could while meeting environmental goals. The key to success is to shift the paradigm from energy security to energy sustainability.

In closing this chapter on the rationale for using nuclear power as a major source for clean, affordable, safe energy, I feel compelled to share with you the thoughts of James Lovelock. He is one of the world's foremost environmentalists and self-proclaimed "green." He is also the author of the environmental theory "Gaia," the hypothesis of which is that planet earth keeps itself regulated and thriving by the actions of the living things themselves. His insights on nuclear power closely track many of the arguments I have shared with you in this chapter about "why nuclear."

James Lovelock has issued a "call to arms" in favor of nuclear power. He believes nuclear power is the only *green* option we have available *now*. Proclaiming that climate changes pose a greater threat to civilization than anything yet faced (including terrorism), Lovelock argues that there is almost no time left to reverse the tide of environmental disaster:

> So what should we do? But with six billion [world population], and growing, few options remain; we can not continue drawing energy from fossil fuels and there is no chance that renewables, wind, tide and water power can provide enough

energy and in time. If we had 50 years or more we might make these our main sources. But we do not have 50 years; the Earth is already so disabled by the insidious poison of greenhouse gasses that even if we stop all fossil fuel burning immediately, the consequences of what we have already done will last for 1,000 years. Every year that we continue burning carbon makes it worse for our descendants and for civilization.

By all means, let us use the small input from renewables sensibly, but only one immediately available source does not cause global warming and that is nuclear energy. True, burning natural gas instead of coal or oil releases only half as much carbon dioxide, but unburnt gas is 25 times as potent a greenhouse agent as is carbon dioxide. Even a small leakage would neutralize the advantage of gas.

Opposition to nuclear energy is based on irrational fear led by Hollywood-style fiction, the Green lobbies and the media. These fears are unjustified, and nuclear energy from its start in 1952 has proved to be the safest of all energy sources. We must stop fretting over the minute statistical risks of cancer from chemicals or radiation. Nearly one-third of us will die of cancer anyway, mainly because we breathe air laden with that all pervasive carcinogen, oxygen. If we fail to concentrate our minds on the real danger, which is global warming, we may die even sooner, as did more than 20,000 unfortunates from overheating in Europe last summer.

But I am a Green and I entreat my friends in the movement to drop their wrongheaded objection to nuclear energy. . . . We have no time to experiment with visionary energy sources; civilization is in imminent danger and has to use nuclear—the one safe, available, energy source—now or suffer the pain soon to be inflicted by our outraged planet.[63]

Roadmap for the Future

My Nuclear Vision

I'm a strong champion of a vision wherein nuclear materials play a vital role in ensuring reliable energy supplies for the world. In my vision, nuclear energy can be one of the tools by which we help the world obtain clean sources of reliable energy—energy that can be used to dramatically advance standards of living around the world. If we are to ameliorate some of the causes of war, proliferation, and terrorism—such as poverty—we must ensure that other countries have access to clean energy supplies to drive their engines of economic development. The seeds of unrest and terrorism will be far less fertile as the standards of living of all peoples are raised toward our own.

Nuclear energy offers solutions to some of the crucial issues facing the world today. Nuclear technologies can provide power for emerging economies while protecting our environment; supply energy for the hydrogen economy while conserving fossil fuels for other uses in an environmentally friendly manner; desalinate water for the world's growing population; improve people's health through application of nuclear medicine; add to the safety of our food supply through irradiation; and improve the safety of industrial products and techniques.

I strongly support the continued development of nuclear power both here and abroad. New nuclear power plants offer emission-free power sources, help maintain diversity of fuel supply, enhance energy security, meet growing electricity demand, protect consumers against the volatility in the prices of home heating oil and natural gas, and provide an economically competitive supply of energy for our growing economy. I would also argue that nuclear power has the potential to break the proliferation cycle by burning weapons-useable materials in commercial reactors, turning atoms for weapons into atoms for energy.

Nuclear power has what it takes. Dr. Denis Beller's acronym CARESS sums up nuclear power's attributes—clean, affordable, reliable, environmentally acceptable, safe and secure, and sustainable.[1]

Bipartisan Leadership Committed to a Diversified, Sustainable Energy Policy

We are in an era of increasingly tight energy supplies, just when the economy is poised to take off again, necessitating even more energy. Instead of embracing new opportunities for clean, sustainable sources of fuel, people are going BANANAS! By BANANAS I mean "Building Absolutely Nothing Anywhere Near Anyone." People have taken the NIMBY ("Not In My Backyard") syndrome to an extreme. Without doubt, if we are going to improve the energy situation in our country, and around the world, we must have an energy policy that balances energy supply, environmental protection, economic growth, and national security. We cannot let arguments based on the rhetoric of environmentalism or security overshadow realistic assessments of our energy needs and abilities. We must pursue a balanced approach. Going BANANAS is a shortsighted, illogical response. The balanced approach is to take a step forward and diversify our energy portfolio with clean supplies of electricity from nuclear power.

While the current Bush administration has made great strides in understanding the need for a national energy policy and in supporting key government initiatives, nuclear power, which absolutely must be part of a diversified energy strategy, will only move forward under a sustained, politically stable leadership effort. It must be bipartisan and consistent, regardless of the leadership or party affiliation of the congressional or executive branches of the U.S. government. The nuclear utility industry and investment community must have faith that the U.S. government's energy policy will not change in midstream. The stakes are too high to be subject to political whim and environmental policy du jour.

The country cannot afford leadership that equates reliance solely on renewables or conservation as the answer to fueling American and global economic growth. Yes, we need those options, but we also need clean sources of dependable baseload electricity from nuclear power and clean coal facilities. We cannot pursue an energy policy that relies on warm winters and cool summers to keep our energy supplies and prices stable. Our economy is growing; hence, affordable, sustainable energy supplies are essential to our continued economic success.

What the country needs is a Kennedyesque initiative, similar to President Kennedy's promise that America would land on the moon. The president of the United States should commit America to take every step necessary to become energy-independent, protect the environment, and provide the engine for economic growth. We need a long-term national commitment to develop all energy sources, to conserve energy, to pursue advanced nuclear fuel cycle power plants, and to move to a hydrogen fuel economy. The centerpiece of American energy policy must be a balanced energy strategy based on advanced nuclear power technologies, clean coal technology, the hydrogen economy, greater efficiency in our cars and consumer products, and increased usage of renewables. Congress must provide real support for a comprehensive, national plan to put the United States on the path to sustainability. The Energy Policy Act is the right first step, as is the president's hydrogen initiative. It has been nearly a decade since the United States passed an energy policy bill. Nearly three years and three hundred hours of hard work and debate have produced the bipartisan Energy Policy Act, and it must be enacted.

The heartening fact is that the climate for nuclear energy in America is changing, the pendulum of public discourse is swinging back in favor of nuclear energy. How else to explain the emergence of nuclear industry consortia exploring the waters in advance of ordering a new nuclear plant? MIT, one of America's most respected academic institutions, suggests that nuclear power will be needed to protect the environment from greenhouse gases. Public opinion polls show increasing public support for nuclear power. I believe we are witnessing the beginning of a new era for nuclear power.

Support New Nuclear Plant Orders

The first step toward a sustainable energy plan is a commitment to rejuvenate nuclear power's future by reversing the quarter-century dearth of new plant construction. We simply must see new domestic plants constructed. It does us little good to compliment ourselves on our foresight in developing and licensing advanced reactor designs—U.S. designs that are providing reliable power elsewhere in the world today—when we have none of these reactor designs under constructions here. It is my priority to break this mold and bring the federal government and nuclear industry together to devise acceptable, realistic, and meaningful incentives for construction of the next generation of nuclear power plants in the United States. I support federal assistance to address the first-of-a-kind expenses associated with engineering and certification of these new plants. The energy legislation I have tirelessly supported included an energy production tax credit of 1.8 cents/kWh, an idea that should remain on the table in our quest to level the playing field and encourage the construction of new nuclear plants.

I believe that once the new plants are built, the industry will show the public that nuclear power plants can be built economically and operated safely, consistent with the public's demands. With those new plants, the industry will be able to convince the investment community that new plant construction is a solid investment opportunity, not one to be shunned.

The industry is taking the necessary measures to be ready to make a decision on ordering a new nuclear plant when the market conditions are ripe. As the economy recovers, the need for new electricity supplies will increase both nationally and regionally, and the market will absorb the overbuilding of new gas plants in the recent past. At that point in time, utilities will make decisions whether to build new coal, new gas, or new nuclear plants. The three nuclear industry consortia described in chapter 4 indicate to me that nuclear power plants may very well be the next choice for new baseload electricity supply in the next five years. Presently they are in the process of testing the nuclear regulatory process and taking care of predecisional items such as site banking or design precertifications. Without concerns over excessive regulatory risks, decisions can be made on a level playing field.

Nothing Is More Important for the Future of Nuclear Power than the Present

A key reason behind the renaissance of nuclear power—increasing political support for nuclear programs and increased public acceptance—is the fact that the current fleet of 103 operating reactors is safe, reliable, and economical. The nuclear industry deserves a lot of credit for taking the steps that have led to increased plant output, rising efficiency factors, an excellent safety record, decreased generating costs, and increased reliability factors. Since 1990, there has been an increase in U.S. nuclear power electricity output equivalent to the amount that would have been generated by twenty-five new 1,000 MWe nuclear plants.

The nuclear industry is undertaking efforts to reduce the capital and operating costs of new nuclear power plants. The improved regulatory process should reduce siting and licensing times thereby significantly reducing the costs of capital. The industry will take advantage of standardized designs to keep the risks in the licensing process to a minimum, as well as to reduce costs in training, operation, and maintenance. Advances in modular construction techniques should improve plant construction times, and—in the nuclear business—time is money.

Nuclear power will stand or fall on its record.

Reach Out to the American Public

Many Americans do not understand the contributions nuclear power has made to our economic well-being, our national security, the safety of our food supply and other industries, and our health. I have tried to put some of these accomplishments into perspective throughout this book. Congressional, corporate, and government leaders together must work harder to raise the public's awareness of the absolute necessity of nuclear power as a vital component in the energy supply mix, not only for our country but for the world. Nuclear's benefits must be better explained to our citizens.

To this end, we need to increase the public's knowledge of the relatively low risks of nuclear power compared to the tremendous advantages it brings. People need to know just how safe nuclear power is before they can embrace it for the future. With the naval reactor program and the history of the nuclear Navy as benchmarks, the American public can begin to understand this for themselves. The relative security and health hazards of nuclear power must be better communicated to the public. The public must be made aware of just how safe nuclear power plants are and that our state and local governments are capable to respond to an emergency and protect the public health and safety.

With all the understandable concern over the security of nuclear plants after the terrorist attacks on the World Trade Center and the Pentagon, it is crucial that we show the American public that in the unlikely event of an airliner filled with fuel crashing into a nuclear plant, there would not be a significant health and safety consequence from radiation. While the likelihood of such an occurrence is very small, America will protect the public and there will not be an American Chernobyl.

Politicians, industry officials, research laboratory directors, and military leaders must actively engage members of the public who are legitimately concerned about nuclear power and who need more information about its public health benefits, its superior economics, and what its risks are compared to other forms of energy supply. We must show them how the facts and figures from our many years of nuclear power operations demonstrate it can be and is safe. In dealing with the opposition that ignores the realm of facts and real science, I think for too long we have become entangled in the impossible challenge of proving a negative—that is, proving that nuclear power is not a risky business. Given the stellar record of the nuclear Navy and the U.S. commercial nuclear power industry, it's time to put the onus of responsibility on "the other side" to prove what they claim—that harm has been done. I don't believe they can.

Industry's role in improving public acceptance of nuclear power begins with the communities they serve today. Being a good, open, honest neighbor gives the populations surrounding current nuclear facilities the assurance and confidence to move forward with more nuclear power in the future. I believe, too, that industry can and should talk more often about the advantages of nuclear power. Industry leaders should be part of an outreach and education effort to the American public. The industry's associations do an excellent job in this regard, but individual leaders should also take a higher profile and a more aggressive role in educating the public.

Since the breakup of the Atomic Energy Commission, no one government agency has been given the dedicated task of educating the public about nuclear power. This has resulted in a huge gap in the public's understanding of the relative risks of nuclear power as compared to other forms of energy. Vice President Cheney, who is clearly versed on the importance of nuclear energy, has called for a public education campaign. We need to develop better ways of communicating the safety and health hazards of all energy technologies relative to each other. Until the public is convinced about the merits on balance of nuclear power, all the messages of clean and green nuclear power will fall on deaf ears.

The nuclear energy industry has recently conducted a reasonably effective campaign to help our citizens understand the role that nuclear technologies play. But I also think that the record of our Navy should be better publicized.

I want to stress again that the U.S. naval nuclear program holds important lessons for better understanding of nuclear power. The *Nautilus*, our first nuclear-powered submarine, was launched in 1954. Since then, the Navy has launched over two hundred nuclear-powered ships, and eighty-two are currently in operation. Recently, the Navy was operating slightly over one hundred reactors, about the same number as those operating in civilian power stations across the country.

The Navy's safety record is exemplary and bears repeating. Our nuclear ships are welcomed into over 150 ports in over 50 countries. Our nuclear-powered ships have traveled over 128 million miles without serious incidents. Further, the Navy has commissioned thirty-three new reactors in the past decade, in stark contrast to the record of civilian power. And Navy reactors have about twice the operational hours of our civilian systems. This is a record of safety, an achievement that I believe the public needs to better understand. It is a solid basis upon which public confidence in nuclear power can be built.[2]

Challenge the World to Strengthen the Rulebook on Nonproliferation

Proliferation is not driven by the spread of nuclear technology, but by political calculations as to the worth of possessing a nuclear weapon and the risks of obtaining one. Approximately two dozen countries have the technology to make nuclear weapons while less than ten have a nuclear bomb. The technology genie is out of the bottle, but that does not mean that nuclear weapons will continue to proliferate.

The safeguarded nuclear fuel cycle has not been the preferred route to nuclear-weapons procurement. Nevertheless, access to nuclear materials, either for a radiological/dirty bomb or for a nuclear weapon, must be curtailed. Meeting the nonproliferation challenge will depend on continued efforts to strengthen the international nonproliferation regime so as to reduce the amount of weapons-useable materials to the greatest extent possible. The regime can be improved dramatically through technology advances in fuel cycles and safety and safeguards. The political dimension must not be overlooked, as the regime can be improved by better IAEA enforcement of the rules and expanded authority for the agency.

It is the United States' responsibility as a world leader to stay engaged in supporting the nonproliferation regime, and the starting point is leading the way in developing proliferation-resistant fuel cycle technologies. The United States cannot hold itself aloof and expect other nations to follow, as President Carter attempted to do. History has proven him wrong. Eighteen countries have developed nuclear fuel cycle capabilities.[3]

Furthermore, the United States should actively become involved in developing alternative institutional arrangements for providing sensitive fuel cycle services such as enrichment or reprocessing—if we choose that path—to other countries, both as incentives to produce clean electricity and as disincentives to countries developing indigenous proliferation-sensitive technologies.

We should work through international organizations like the IAEA to craft global approaches to fuel production and waste handling in ways that minimize proliferation concerns—we just require the will to work through and with the international community to make them a reality.

Our nation will, I trust, proceed with development of nuclear power here at home. But my vision, like President Eisenhower's, doesn't stop at our own borders. We must help provide developing nations with the energy resources so that they too can grow and prosper. The seeds of unrest and terrorism will be far less fertile as the standards of living of all peoples are raised toward our own.

As we assist these nations, we should suggest that they focus their attentions on clean energy sources to avoid the environmental problems that many developed nations like the United States are experiencing with past energy sources. Nuclear should be one of the clean technologies we offer to them.

Now obviously, many of those developing nations do not have the necessary infrastructure to produce and safeguard nuclear materials or to design new reactors. But we can and should help them in specific ways, such as providing small sealed reactors, along with full assurances from the developed nations to guarantee to provide all their life-cycle fuel services. Those same life-cycle fuel services should be offered to many other nations for large reactors as well.

Advanced Nuclear Technology Development

The United States has been a world leader in both the policy and technical aspects of nuclear energy. This nation operates more nuclear power plants than any other country, and most of the world's operating nuclear power plants are based on U.S. LWR technology. Given:

- the projected growth in global energy demand as developing nations industrialize;
- our strategic interests in addressing nuclear nonproliferation, nuclear safety, economic competitiveness, and potential global climate change; and
- our need to satisfy growing domestic needs for energy in an environmentally responsible manner,

the United States must regain its scientific and technological leadership in nuclear energy. This leadership will provide the United States a key "seat at the table" at ongoing international discussions regarding the future implementation of nuclear technologies, nuclear nonproliferation, nuclear safety, and many other issues important to U.S. policy objectives.

The United States' leadership in nuclear technology is a vital component of our nation's foreign policy. American prominence in nuclear technology enables the United States to exercise considerable influence on the manner in which nuclear technologies are applied worldwide. Strategies to prevent nuclear materials proliferation, nuclear safety practices, and safeguards policy are directly influenced by U.S. technical leadership in the international community.

The United States has lost its lead in the development of nuclear fuel cycle technologies since the Carter administration set the nation on a course of an open fuel cycle. We must consequently redouble our efforts in the R&D area and take the lead in developing advanced fuel cycle technologies and next-generation nuclear power systems. Toward this end, our Advanced Fuel Cycle Initiative and the Generation IV Nuclear Energy System Initiative are key programs to enable the United States to take its place in the dialogue on the future of nuclear power.

Congress must support major research, demonstration, and deployment reactor programs to demonstrate a new generation of ultrasafe, ultraefficient reactors that minimize waste production and proliferation concerns. We must find ways via new fuel cycle technologies to more wisely use our uranium resources. Whether one or all of the resource conservation, proliferation, or waste disposal concerns drive our decisions regarding the fuel cycle, America must be prepared in the next ten years to make choices based on solid research and development. This effort must be robustly supported today and sustained into the foreseeable future.

In keeping with this concept, we must make progress on developing advanced strategies for spent nuclear fuel to allow us to benefit from its energy value and at the same time furthering our nonproliferation goals. Additionally, we must undertake R&D of technologies that would allow us to reduce the toxicity of the waste products that must ultimately be buried. And as I said, the barriers to opening our first repository are political, not technical. We are already pursuing processing and transmutation technologies that will eventually allow us to achieve our energy *and* national security goals.

Progress toward a hydrogen economy is another major goal. The opportunity to convert our transportation system to hydrogen fuel offers immense benefits, but only if we can produce that hydrogen cleanly and economically. As we all know, we can't mine or drill for hydrogen—we have to produce it. That requires energy. I support alternative production approaches, from solar to nuclear to biological.

Wider use of advanced nuclear power technologies should be embraced as a means for fueling economies, improving international security, and improving the environment.

New Approaches to Nuclear Waste Management

In the area of nuclear waste policy, I'm convinced that our citizens demand better solutions than we have today. Now we plan to simply bury and forget about our spent fuel—never mind that it retains an immense store of energy or that its constituents are highly toxic.

Earlier in this book I mentioned the expensive black hole that we call Yucca Mountain, the Nevada site that may some day become a permanent storage facility for nuclear waste generated by the nation's power plants. So far, we have spent $4.8 billion on the initiative. The implacable opposition of Nevada politicians, especially Senate Democratic Whip Senator Harry Reid, has resulted in an economic boon to Nevada, thanks to all that federal spending, but has left the nation as a whole with delay after delay to show for it. Clearly, our policy has gone astray.

The United States must stay the course toward licensing and constructing the country's first waste repository. We need continual, meaningful progress in the development of the Yucca Mountain repository. I hope that the courts will expeditiously and fairly resolve the legal challenges brought by the state of Nevada to the decision to site a repository at Yucca Mountain and to the process itself. We are already late in opening the repository. Time taken now in carefully resolving the legal issues in the proper manner will not set the program back. On the contrary, taking the time to do it right will ultimately make for a better outcome. In the meantime, I hope that the state of Nevada will end its legally obstructionist tactics to become engaged constructively in the process of developing this repository. The people of Nevada will ultimately benefit from this approach and there are indications that the people of Nevada increasingly feel that way.

Progress on the waste management front is embedded in stable political and regulatory affairs that I flagged as key factors in the resurgence of nuclear power. It also involves improved public perceptions that progress has indeed been made and that our engineers will be able to develop a repository design that will allow for safe waste disposal. Energy leaders owe it to the American people to focus on the positive, and not on the divisive road advocated by nuclear power opponents who would shut down plants until a repository opens.

As I study the problem, I am persuaded that the nation cannot put all its waste in one basket, so to speak. We need a policy of interim storage. Ideally, such a site would be one that has already had great experience with nuclear material, licensed to handle such material at all levels of intensity, and which has a highly skilled workforce in place. Complicating any change in policy on nuclear waste management is the fact that the nation's utility companies— and their ratepayers—finance the program. Such companies have paid their billions in a special federal fund, but they still have to store the waste themselves until Yucca Mountain becomes a reality.

I've become convinced that interim storage and continued work on Yucca Mountain cannot be the complete answer. If our nation's only long-term solution for spent fuel is to put it in a permanent repository, we have signed the death knell for nuclear power. Emplacement of spent fuel in a repository inspires real fears among a site's neighbors, as Nevada, albeit with a lot of help from antinuclear groups, has demonstrated. And any expansion of nuclear energy means we will generate vastly more spent fuel than can be placed in a repository like Yucca Mountain, meaning that nuclear energy will literally choke on its inability to handle its wastes. This is the goal of antinuclear groups.

My studies have convinced me that there are far smarter ways of handling spent fuel than dumping them in a hole in the ground, as discussed in detail in chapter 9. Advanced technologies hold the promise of a massive reduction in volume and toxicity of the final waste products with recovery of the 95 percent of the energy content that is in our spent fuel today; but all approaches would still require that some final small quantities of wastes go into a repository. With interim storage, we buy time for serious research on alternative strategies for spent fuel. At the same time we should continue scientific studies toward realization of our first repository for high-level waste. And when the first repository opens, we need to be sure that the fuel is retrievable for many decades while research on better solutions advances.

I've championed the study of these "smarter" solutions over the last few years. I do not know whether the promises of accelerator-driven or fast reactor-transmutation technologies will be deployable. However, they have enough promise that we must pursue them at the R&D, and potentially, at the RD&D levels. Their development should take place in tandem with the DOE's efforts to open a geologic repository and to get the first new order for a new nuclear power plant. There is adequate time to develop these new waste management strategies.

In the 109th Congress, beginning in January 2005, I will seriously consider legislation beginning a process to designate an interim storage site—one that the nation can use for fifty to seventy-five years. Yes, we need a permanent storage facility, but the nation is now held hostage by current reliance on Yucca Mountain.

Rebuild Our Nuclear Infrastructure and Human Resources

We need a renewal of nuclear energy initiatives in this country. One component of making that happen is addressing the "graying" of the technical expertise needed to support nuclear technologies. We just are not training enough new specialists to support a range of critical initiatives in this country that require nuclear engineering knowledge. We need stronger university programs with more students challenged and motivated to master these technologies to contribute to the future workforce. Legislation is needed to support university nuclear science, health physics, and engineering programs—to fund research, training, fellowship programs, recruitment, and retention of excellent professors, improvements in the universities' research and training reactors, and fuel assistance for university research reactors.

Of equal priority is the need to help our national laboratories rebuild their technical expertise and infrastructure, which suffered from neglect—a neglect that was especially acute during the Clinton administration years. This speaks to the need for a broad-based R&D program dedicated to making the United States self-sufficient in terms of its energy needs. Absent this consensus on our energy policy goals, our laboratories will flounder and our spending priorities will not be focused on the sustainability goal. The United States will be able to regain its technology and leadership status in the nuclear arena only if we dedicate adequate resources and direction to the treasure of our national laboratories.

Stabilize the Regulatory Framework

We have recognized the huge costs incurred by our nation during the turbulent regulatory climate of the 1970s and 1980s. The goalposts were constantly moving, the process itself was flawed, and the industry and regulators distrusted each other. Today we realize the importance of basing decisions on risk-informed, performance-based regulations. New processes are in place to ensure that licensing a nuclear power plant does not drag on indefinitely. Contributing to this progress is one-stop licensing, precertification of new reactor designs, and site banking. The focus is appropriately on ensuring the public's safety as well as the viability of the developer's capital investment. The next few years will be crucial in determining whether the regulatory process facilitates rather than hinders the future development of nuclear power in the United States.

We may never be able to calculate the true costs of overregulating, but common sense tells us that regulations based on false conjecture or even psuedoscience will tax the American public unnecessarily. That's how I feel about basing our radiation protection standards on the LNT model. We owe it to our country to develop tough but fair nuclear regulations. This is why I have relentlessly prodded our government to do the research documenting molecular and cellular responses to low doses of radiation. The National Academy of Sciences is now poised to make recommendations based on the scientific information gathered by the DOE for a risk model for exposures for ionizing radiation. Their report will come none too soon and I believe it will help to put us on the path to sound, risk-based nuclear radiation standards.

The Promise of the New Nuclear Paradigm

The kernel of the ideas in this book began with my speech at Harvard on October 31, 1997. I made a pledge to all American citizens that I would exert leadership to find answers to what went wrong and to what needs to be done to fix the problems of nuclear power. I said:

> I hope in these remarks that I have succeeded in raising your awareness of the opportunities that our nation should be seizing to secure a better future for our citizens through careful reevaluation of many ill-conceived fears, policies and decisions that have seriously constrained our use of nuclear technologies.
>
> Today I announce my intention to lead a new dialogue with serious discussion about the full range of nuclear technologies. I intend to provide national leadership to overcome barriers.
>
> While some may continue to lament that the nuclear genie is out of his proverbial bottle, I'm ready to focus on harnessing that genie as effectively and fully as possible, for the largest set of benefits for our citizens.
>
> I challenge all of you to join me in this dialogue to help secure these benefits.

I have always believed that man can solve his problems with the proper determination and effort. In his "Atoms for Peace" speech before the United Nations General Assembly on December 8, 1953, President Eisenhower dedicated the United States to the task of solving the nuclear dilemma. He said: "To the making of these fateful decisions, the United States pledges before you, and therefore before the world, its determination to help solve the fearful atomic dilemma—to devote its entire heart and mind to finding the way by which the miraculous inventiveness of man shall not be dedicated to his death, but consecrated to his life."

My fervent desire is that I help revive America's pledge to the world to bring the awesome benefits of nuclear power to the betterment of all people.

Epilogue

Mid-spring 2004 reveals Washington, D.C., at its most beautiful. Outside the Capitol, the first of millions of tourists wander through the azaleas and dogwoods and oaks. To the casual observer, the world moves in an easy and reassuring pace.

Inside the Capitol, by contrast, everything has come to a halt.

The Senate, especially, finds it difficult to move any legislation. Under Senate rules, an obstructionist minority can stop almost any bill, no matter how important. As I write, the Democratic minority has bottled up one reform measure after another. One victim of this "gridlock" is the Energy Policy Act, the bill that I shepherded through the Energy and Natural Resources Committee in May 2003, and which has since consumed more Senate floor time than any other bill in the last forty years.

Ironically, late in 2003, a bipartisan majority in the Senate voted in favor of the bill, 57 to 40. But the vote came on a procedural motion that, under Senate rules, required sixty senators to vote in favor. So, although a majority of the Senate spoke, its voice was throttled by the Democratic leadership's strategy of prolonged debate and obstructionism. Since then, the bill has been in a kind of legislative limbo, despite the tireless work of Senate Majority Leader Senator Bill Frist to move it to a vote. It may happen that Senator Frist and I will be able to shepherd a final version of the bill through the Senate and send it to President Bush for his signature, but no predictions ever seem safe in the chaos of the 108th Congress.

Why is it so important to pass the energy legislation? I think of the events in the past decade that the energy policy bill responds to, events I have noted in this book. The reasons are compelling:

- natural gas prices continue to soar amid warnings of global shortages within the decade, making the Alaska Natural Gas Pipeline more important than ever;

- using more renewables in gasoline may help contain the price of gas at the pump, a price that has hit $2 per gallon throughout the nation;
- if coal is to continue to produce a large percentage of our energy needs in the future, we must expand clean coal technology efforts so that pollution can be better controlled, just as the Bush administration has called for new mercury emissions control from coal-fired generators;
- wind, solar, and to a lesser extent other "alternative energy sources" cannot be expected to provide a large percentage of our baseload energy needs in the foreseeable future, but are still needed despite the protests of citizens who dislike the environmental impact of such sources;
- mandatory reliability standards, as shown to be more critical than ever after the 2003 blackout that hit the northeastern United States and cost the economy an estimated $40 billion must be strengthened; and
- as this book argues, incentives are needed for investment in advanced nuclear energy plants.

In short, each provision of the proposed bill directly addresses critical and immediate energy policy problems. Each has been prompted by recent crises and near-crises. Each is relevant and necessary.

Yet, Congress cannot act.

Congress cannot blame its weakness on the American public. The American public is very clear: energy dependence on overseas sources of oil is bad, high gasoline prices are bad, natural-gas price increases are bad, thinking that solar or wind will solve all our problems is wrong, and clean air and water are good. For too long, a combination of well-meaning dilettantes, "utopians" longing for a world in which all good things come without cost, and hard-eyed political partisans have been able to stymie the Senate. We need alternative energy sources; we need less pollution from coal use; we need energy independence to the extent it is possible. And, the key to all this, in my view, is a reinvigorated nuclear energy strategy for the nation.

The energy bill saga in the 107th and 108th Congress is instructive in many ways.

In the 107th Congress, the Senate Energy and Natural Resources Committee was unable to produce a bill at all. The then majority leader, Senator Tom Daschle, bypassed the committee after a long wait and brought the bill to the Senate floor. After healthy debate, that bill passed the Senate, 84 to 14. The House and Senate, however, could not agree on a compromise bill and the entire effort died with a whimper.

In the 108th Congress, as I have outlined, we were able to produce a bipartisan bill out of committee. The House passed its energy bill by a large, bipartisan margin. In the aftermath of the electricity crisis in California, with rising energy prices, and with joblessness a big political issue, the omens seemed auspicious for Senate passage of our companion measure.

That's when the train came off the tracks. On May 6, 2003, I rose on the Senate floor to start debate on S.14, the Energy Policy Act. I admitted that the committee had deliberately decided not to consider such items as vehicle fleet gas mileage, nor climate change amendments, which are not within the jurisdiction of the committee. But I recognized that we expected amendments on these subjects, and others not strictly germane to the bill, on the Senate floor.

"I sincerely hope that this important legislation does not become wrapped up in partisan delay tactics," I pleaded on the floor. I felt that the two-year-long saga of the energy bill in the 107th Congress and the intensive markup in committee at the start of the 108th Congress coupled with the prospect of full debate on the Senate floor would have given the senators many chances to offer as many amendments as they wished and to get the full counsel of the Senate. My view proved overly optimistic, and wrong.

After literally weeks of desultory debate, in which senators claimed to have amendments but refused to offer them on the floor, the Senate took a couple of key votes. The amendment to increase average vehicle fleet mileage failed overwhelmingly; the climate change amendment proponents were never able to agree on an amendment and that vote never happened. As time when on, more and more senators discovered an endless list of subjects on which they might offer amendments. Never mind that many of these amendments had little to do with real energy policy, and many amendments remained little more than notions in the heads of their proponents. The Senate was truly at an impasse.

At this point, an extraordinary thing happened. Senate Minority Leader Tom Daschle stood up and offered a suggestion. "I make this suggestion for the consideration of the Senate," he said. "Why don't we consider passing the same energy bill that the Senate passed last year."

I was sitting next to Majority Leader Frist on the Senate floor when the Daschle offer was made. Frist looked at me; I looked at him. I immediately made a thumbs-up sign. Frist said that he would like a little time to talk in the Senate cloakroom about the offer. Shortly, Frist and I emerged from the Republican cloakroom and accepted the offer. The full Senate then passed an energy bill—not the bill I had brought to the floor, but the one that the then Democratic majority had passed the year before!

The process wasn't perfect. Hell, it wasn't even pretty. But, warts and all, the Senate had passed an energy bill and we could now go to conference with the House and see if we could hammer out a compromise bill that could eventually go to the president. President Bush had been pounding Congress for more than two years to produce an energy bill and it looked as though we might make it.

During Senate debate, some senators—mostly Republican, surprisingly—had raised objections to provisions in my energy bill that helped nuclear energy get off the ground again. The bill that the Senate eventually passed had no nuclear provisions, so I planned to work hard in conference to salvage my nuclear initiative. I told the Senate and the media forthrightly that I would go to conference with the energy bill that the Senate had given me, but that I would consider all topics open for discussion and would bring back to the Senate a compromise bill that I thought best reflected good policy. As I told my colleagues, I never intended to be bound by the Senate bill that had failed in conference a year earlier.

The Energy Policy Act conference between the House and Senate became a "passion-play" almost at once—but it wasn't the fault of the energy committee participants. At the same time that the energy bill was being negotiated, the House and Senate were conferencing on a Medicare prescription-drug benefit bill. The tax-writing committees of the Senate and House found themselves trying to compromise on Medicare and energy at the same time. Soon, the profound differences between House Ways and Means Committee chairman Bill Thomas and Senate Finance Committee chairman Chuck Grassley came into full, public view.

The media had a field day writing about the meetings and antics between House and Senate tax conferees on Medicare. Long intervals between meetings became commonplace. Every time the two bodies erupted about a Medicare nuance, the energy tax provisions of my energy bill got shunted aside.

At the same time, those of us on the nontax portions of the energy bill conference were having our own difficulties. House Energy and Commerce chairman Billy Tauzin and I got along well; we were both very experienced in legislation, had good staffs, and wanted a bill as soon as possible. But in spite of all of our good will, we found it hard to overcome three policy differences—how to handle ethanol and a substance called methyl tertiary-butyl ether, a fuel oxygenator; what to do about electricity regulation in the almost Balkanized regions of the nation; and, what the overall cost of the bill should be.

Billy and I were able to agree on the nontax portions of the bill, but the tax portions held us hostage. It was hard to agree on ethanol policy, for example, without considering both the authorizing and tax portions of such a policy. We could agree all we wanted in the privacy of my Capitol hideaway, but without agreement on tax policy, we couldn't put together a final compromise package.

After much publicity, presidential intervention, threats, and imprecations, the Medicare bill and our energy bill, taxes and all, came together. We had an energy bill conference agreement. Tauzin took his bill to the House and passed it handsomely. I had hopes that we could do the same. Again, I over-estimated the willingness of the Senate to put the general good above parochial considerations. Ultimately, we could not overcome a combination of trial lawyer interests, interregional hostility, and personal pique.

Representatives from almost every aspect of the nation's energy economy roamed the Senate halls in the days before the vote on the Energy Policy Act. Rural electric cooperatives from every part of the country joined alternative energy advocates, corn and soybean growers, construction labor unions, and hundreds of others in an extraordinary effort to get our bill passed.

As I sat on the Senate floor during the vote, I looked around the chamber. I remembered my colleagues who said that if I just took out the Arctic Na-tional Wildlife Refuge (ANWR) drilling provision, the bill would pass. Well, I had taken ANWR out months earlier. Some of my fellow senators had warned that I had to set reliability standards, or increase subsidies to solar or wind energy. We had done that. Others said that ethanol subsidies were the key. We had historic ethanol policy expansions in the bill.

In the end, it was not enough. A filibuster—an extended debate that is very hard to end—began. It was only one of dozens of filibusters afflicting the Senate in the past decade, but this one was the first ever launched at a major bill that I had authored. I knew we needed sixty votes to kill the filibuster. We got only fifty-seven. With a parliamentary maneuver, Senate Majority Leader Frist was able to keep the bill from absolute defeat. But, it was on life support.

The year 2004 began the way 2003 had ended—more American troops in oil-rich lands, warnings about mercury pollution from coal plants, historically high natural gas and gasoline prices, and deep concerns about job creation within the economy. The external landscape still seemed favorable to our en-ergy bill and its prospects of creating more than 1 million jobs. Behind the scenes we tried tactic after tactic. As of this writing, the Energy Policy Act still remains unpassed, but remains unbeaten as well.

Yet, the prospects for nuclear energy increase despite the setbacks of the energy bill.

This morning I read a *New York Times* article on companies banding to-gether to try to begin the process of applying to build a nuclear power plant. The article started, "Amid growing signs of interest in building nuclear power plants. . . ." As the facts fight hysteria, as good science begins to spread throughout the nation, as the downside of other energy sources begins to be felt, the nuclear option may once again claim its rightful place as a prime part of an intelligent energy policy.

Will I live to see a series of new nuclear power plants built in this land? I'm not sure. But, more important than that, if these plants are built both in America and abroad, my children and grandchildren will live much, much better lives—and the billions of future children in other lands will have a better chance at a decent life. First, however, we must grasp the nuclear option intelligently and courageously.

Appendix A

"A New Nuclear Paradigm"

Earlier this week, I spent substantial time on the subjects of nuclear non-proliferation, the proposed Comprehensive Test Ban Treaty, nuclear waste policies, and nuclear weapons design issues. The forums for these discussions were open and closed hearings of two major subcommittees of the United States Senate, a breakfast where two cabinet secretaries joined ten United States senators, and private discussions with specialists in these fields.

During the week before, I spent time on the question of whether or not a 1,200-foot road should be built in a national monument, a monument whose enabling legislation I authored almost a decade ago.

Without demeaning any person's sense of perspective, I have to note to you today that for every person who attended the nuclear hearings, fifty attended the road hearings. And, for every inch of newspaper coverage the nuclear matters attracted, the road attracted fifty inches.

Strategic national issues just don't command a large audience. In no area has this been more evident during these last twenty-five years than in the critical and interrelated public policy questions involving energy, growth, and the role of nuclear technologies. As we leave the twentieth century, arguably the "American century," and head for a new millennium, we truly need to confront these strategic issues with careful logic and sound science.

We live in the dominant economic, military, and cultural entity in the world. Our principles of government and economics are increasingly becoming the principles of the world.

There are no secrets to our success, and there is no guarantee that, in the coming century, we will be the principal beneficiary of the seeds we have sown. There is competition in the world and serious strategic issues facing the United States cannot be overlooked.

The United States—like the rest of the industrialized world—is aging rapidly as our birth rates decline. Between 1995 and 2030, the number of people in the United States over age sixty-five will double from 34 million to 68 million. Just to maintain our standard of living, we need dramatic increases in productivity as a larger fraction of our population drops out of the workforce.

By 2030, 30 percent of the population of the industrialized nations will be over sixty. The rest of the world—the countries that today are "unindustrialized"—will have only 16 percent of their population over age sixty and will be ready to boom.

As those nations build economies modeled after ours, there will be intense competition for the resources that underpin modern economies.

When it comes to energy, we have a serious, strategic problem. The United States currently consumes 25 percent of the world's energy production. However, developing countries are on track to increase their energy consumption by 48 percent between 1992 and 2010.

The United States currently produces and imports raw energy resources worth over $150 billion per year. Approximately $50 billion of that is imported oil or natural gas. We then process that material into energy feedstocks such as gasoline. Those feedstocks, the energy we consume in our cars, factories, and electric plants, are worth $505 billion per year.

So, while we debate defense policy every year, we don't debate energy policy, even though it already costs us twice as much as our defense, other countries' consumption is growing dramatically, and energy shortages are likely to be a prime driver of future military challenges.

When I came to the Senate a quarter of a century ago, we debated our dependence on foreign sources of energy. We discussed energy independence, but we largely decided not to talk about nuclear policy options in public.

At the same time, the antinuclear movement conducted their campaign in a way that was tremendously appealing to mass media. Scientists, used to the peer-reviewed ways of scientific discourse, were unprepared to counter. They lost the debate.

Serious discussion about the role of nuclear energy in world stability, energy independence, and national security retreated into academia or classified sessions.

Today, it is extraordinarily difficult to conduct a debate on nuclear issues. Usually the only thing produced is nasty political fallout.

I am going to bring back to the marketplace of ideas a more forthright discussion of nuclear policy.

My objective tonight is not to talk about talking about a policy. I am going to make some policy proposals. Tomorrow there are sessions on energy policy and nuclear proliferation. I'll give them something to talk about.

I am going to tell you that we made some bad decisions in the past that we have to change. Then I will tell you about some decisions we need to make now.

First, we need to recognize that the premises underpinning some of our nuclear policy decisions are wrong. In 1977, President Carter halted all U.S. efforts to reprocess spent nuclear fuel and develop mixed-oxide (MOX) fuel for our civilian reactors on the grounds that the plutonium could be diverted and eventually transformed into bombs. He argued that the United States should halt its reprocessing program as an example to other countries in the hope that they would follow suit.

The premise of the decision was wrong. Other countries do not follow the example of the United States if we make a decision that other countries view as economically or technically unsound. France, Great Britain, Japan, and Russia all now have MOX fuel programs.

This failure to address an incorrect premise has harmed our efforts to deal with spent nuclear fuel and the disposition of excess weapons material, as well as our ability to influence international reactor issues.

I'll cite another example. We regulate exposure to low levels of radiation using a so-called linear no-threshold model, the premise of which is that there is no "safe" level of exposure.

Our model forces us to regulate radiation to levels approaching 1 percent of natural background despite the fact that natural background can vary by 50 percent within the United States.

On the other hand, many scientists think that living cells, after millions of years of exposure to naturally occurring radiation, have adapted such that low levels of radiation cause very little if any harm. In fact, there are some studies that suggest exactly the opposite is true—that low doses of radiation may even improve health.

The truth is important. We spend over $5 billion each year to clean contaminated DOE sites to levels below 5 percent of background.

In this year's Energy and Water Appropriations Act, we initiated a ten-year program to understand how radiation affects genomes and cells so that we can really understand how radiation affects living organisms. For the first time, we will develop radiation protection standards that are based on actual risk.

Let me cite another bad decision. You may recall that earlier this year, Hudson Foods recalled 25 million pounds of beef, some of which was contaminated by E. coli. The administration proposed tougher penalties and mandatory recalls that cost millions.

What you may not know is that the E. coli bacteria can be killed by irradiating beef products. The irradiation has no effect on the beef. The FDA does not allow the process to be used on beef, even though it is allowed for poultry, pork, fruit, and vegetables, largely because of opposition from some consumer groups that question its safety.

But there is no scientific evidence of danger. In fact, when the decision is left up to scientists, they opt for irradiation—the food that goes into space with our astronauts is irradiated.

I've talked about bad past decisions that haunt us today. Now I want to talk about decisions we need to make today.

The president has outlined a program to stabilize the U.S. production of carbon dioxide and other greenhouse gases at 1990 levels by some time between 2008 and 2012. Unfortunately, the president's goals are not achievable without seriously impacting our economy.

Our national laboratories have studied the issue. Their report indicates that to get to the president's goals we would have to impose a $50/ton carbon tax. That would result in an increase of 12.5 cents/gallon for gas and 1.5 cents/kWh for electricity—almost a doubling of the current cost of coal or natural gas-generated electricity.

What the president should have said is that *we need nuclear energy to meet his goal.* After all, in 1996, nuclear power plants prevented the emission of 147 MMTC, 2.5 million tons of nitrogen oxides, and 5 million tons of sulfur dioxide. Our electric utilities' emissions of those greenhouse gases were 25 percent lower than they would have been if fossil fuels had been used instead of nuclear energy.

Ironically, the technology we are relying on to achieve these results is over twenty years old. We have developed the next generation of nuclear power plants—which have been certified by the NRC and are now being sold overseas. They are even safer than our current models. Better yet, we have technologies under development like passively safe reactors, lead-bismuth reactors, and advanced liquid metal reactors that generate less waste and are proliferation resistant.

An excellent report by Dr. John Holdren for the President's Committee of Advisors on Science and Technology calls for a sharply enhanced national effort. It urges a "properly focused R&D effort to see if the problems plaguing fission energy can be overcome—economics, safety, waste, and proliferation." I have long urged the conclusion of this report—that we dramatically increase spending in these areas for reasons ranging from reactor safety to nonproliferation.

I have not overlooked that nuclear waste issues loom as a roadblock to increased nuclear utilization. I will return to that subject.

For now, let me turn from nuclear power to nuclear weapons issues.

Our current stockpile is set by bilateral agreements with Russia. Bilateral agreements make sense if we are certain who our future nuclear adversaries will be and are useful to force a transparent build-down within Russia. But I will warn you that our next nuclear adversary may not be Russia—we do not want to find ourselves limited by a treaty with Russia in a conflict with another entity.

We need to decide what stockpile levels we really need for our own best interests to deal with any future adversary.

For that reason, I suggest that, within the limits imposed by START II, the United States move away from further treaty imposed limitations and move to what I call a "threat-based stockpile."

Based upon the threat I perceive right now, I think our stockpile could be reduced. We need to challenge our military planners to identify the minimum necessary stockpile size.

At the same time, as our stockpile is reduced and we are precluded from testing, we have to increase our confidence in the integrity of the remaining stockpile and our ability to reconstitute if the threat changes. Programs like science-based stockpile stewardship must be nurtured and supported carefully.

As we seriously review stockpile size, we should also consider stepping back from the nuclear cliff by de-alerting and carefully reexamining the necessity of the ground-based leg of the nuclear triad.

Costs certainly aren't the primary driver for our stockpile size, but if some of the actions I've discussed were taken, I'd bet that as a bonus we'd see major budget savings. Now we spend about $30 billion each year supporting the triad.

Earlier I discussed the need to revisit some incorrect premises that caused us to make bad decisions in the past. I said that one of them, regarding reprocessing and MOX fuel, is hamstringing our efforts to permanently dismantle nuclear weapons.

The dismantlement of tens of thousands of nuclear weapons in Russia and the United States has left both countries with large inventories of perfectly machined classified components that could allow each country to rapidly rebuild its nuclear arsenals.

Both countries should set a goal of converting those excess inventories into nonweapon shapes as quickly as possible. The more permanent those transformations and the more verification that can accompany the conversion of that material, the better.

Technical solutions exist. Pits can be transformed into nonweapon shapes and weapon material can be burned in reactors as MOX fuel, which by the way is what the National Academy of Sciences has recommended. However, the proposal to dispose of weapons plutonium as MOX runs into that old premise that MOX is bad despite its widespread use by our allies.

MOX *is* the best technical solution. I challenge you to develop a proposal that brings the economics of the MOX fuel cycle together with the need to dispose of weapons-grade plutonium. Ideally, incentives can be developed to speed Russian materials conversion while reducing the cost of the U.S. effort. The idea for the U.S.–Russian HEU Agreement originated at MIT, and I know that Harvard does not like to be upstaged.

I said earlier that I would not advocate increased use of nuclear energy and ignore the nuclear waste problem. The path we've been following on Yucca Mountain sure isn't leading anywhere very fast. I'm about ready to reexamine the whole premise for Yucca Mountain.

We're on a course to bury all our spent nuclear fuel, despite the fact that a spent nuclear fuel rod still has 60–75 percent of its energy content—and despite the fact that Nevadans need to be convinced that the material will not create a hazard for over one hundred thousand years.

Our decision to ban reprocessing forced us to a repository solution. Meanwhile, many other nations think it is dumb to just bury the energy-rich spent fuel and are reprocessing.

I propose we go somewhere between reprocessing and permanent disposal by using interim storage to keep our options open. Incidentally, sixty-five senators agreed with the importance of interim storage, but the administration has only threatened to veto any such progress and has shown no willingness to discuss alternatives.

Let me highlight one attractive option. A group from several of our largest companies, using technologies developed at three of our national laboratories and from Russian institutes and their nuclear navy, discussed with me an approach to use that waste for electrical generation. They use an accelerator, not a reactor, so there is never any critical assembly. There is minimal processing, but carefully done so that weapons-grade materials are never separated out and so that international verification can be used. And when they get done, only a little material goes into a repository, but now the half-lives are changed so that it's a hazard for perhaps three hundred years—a far cry from one hundred thousand years. It sure would be easier to get acceptance of a three-hundred-year rather than a one-hundred-thousand-year hazard, especially when the three-hundred-year case is also providing a source of clean electricity. This approach, called accelerator transmutation of waste, is an area I want to see investigated aggressively.

I still haven't touched on all the issues imbedded in maximizing our nation's benefit from nuclear technologies, and I can't do that without a much longer speech.

For example, I haven't discussed the increasingly desperate need in the country for low-level waste facilities like Ward Valley in California. In California, important medical and research procedures are at risk because the administration continues to block the state government from fulfilling their responsibilities to care for low-level waste.

And I haven't touched on the tremendous window of opportunity that we now have in the former Soviet Union to expand programs that protect fissile material from moving onto the black market or to shift the activities of former Soviet weapons scientists onto commercial projects. Along with Senators Nunn and Lugar, I've led the charge for these programs. Those are programs directly in our national interest. I know that some national leaders still think of these programs as foreign aid; I believe they are sadly mistaken.

We are realizing some of the benefits of nuclear technologies today, but only a fraction of what we could realize.

Nuclear weapons, for all their horror, brought to an end fifty years of worldwide wars in which 60 million people died.

Nuclear power is providing about 20 percent of our electricity needs now and many of our citizens enjoy healthier, longer lives through improved medical procedures that depend on nuclear processes.

But we aren't tapping the full potential of the nucleus for additional benefits. In the process, we are short-changing our citizens.

I hope that in these remarks I have succeeded in raising your awareness of the opportunities that our nation should be seizing to secure a better future for our citizens through careful reevaluation of many ill-conceived fears, policies, and decisions that have seriously constrained our use of nuclear technologies.

Today I announce my intention to lead a new dialogue with serious discussion about the full range of nuclear technologies. I intend to provide national leadership to overcome barriers.

While some may continue to lament that the nuclear genie is out of his proverbial bottle, I'm ready to focus on harnessing that genie as effectively and fully as possible, for the largest set of benefits for our citizens.

I challenge all of you to join me in this dialogue to help secure these benefits.

Appendix B

"Atoms for Peace— 50th Anniversary"

I commend the leadership of the Woodrow Wilson International Center for Scholars, the College of William and Mary, and the Los Alamos National Laboratory for organizing this conference on the occasion of the fiftieth anniversary of Atoms for Peace.

President Eisenhower demonstrated remarkable vision in presenting his remarks to the United Nations fifty years ago. At that time the framework of the challenge confronting the world with nuclear technologies was only dimly visible. But with that speech, he launched this nation and the world on a quest to harness the atom solely for purposes of peace—a quest that remains unmet today.

At the time of his speech, our nation had conducted forty-two test explosions and used two weapons to end the catastrophic Second World War. The Soviet Union had demonstrated their nuclear weapons capability. But the dimensions of the Cold War, which President Eisenhower hoped to avoid, were only vaguely defined. Nevertheless, his words ring true today:

"Let no one think that the expenditure of vast sums for weapons and systems of defense can guarantee absolute safety for the cities and the citizens of any nation. The awful arithmetic of the atomic bomb does not permit of such an easy solution."

The president further noted that the ability of the United States to lay waste to an aggressor would be a hollow victory indeed—hardly befitting the founding principles and ideals of this nation. He noted that such reasoning would be: "to accept helplessly the probability of civilization destroyed—the annihilation of the irreplaceable heritage of mankind handed down to us generation from generation—and the condemnation of mankind to begin all over again the age old struggle upward from savagery toward decency and right and justice."

In some ways, we've come a long way from Ike's words. But in other ways, we still have a long way to go.

Frustration over our slow progress forty-four years after the Atoms for Peace speech led me to speak at Harvard in 1997 and issue a challenge to the nation's leaders to work toward realizing the promise of nuclear technologies to benefit all of mankind.

Indeed, much of my energies over the last decade have been devoted to realizing more of the vision of Atoms for Peace, from both the military and civilian perspectives. I've frequently noted that the citizens of the world will fully realize the benefits of the atom only when the nations of the world have fully controlled and mitigated the military threats of these technologies.

I've championed many initiatives on the military side of this equation, from Cooperative Threat Reduction, to the Material Protection Control and Accounting program, the HEU deal, and the Plutonium Disposition program.

We've seen good progress in some of these initiatives. Certainly the Nunn–Lugar program has accomplished a great deal. Significant quantities of fissile materials are under better control today. There is less HEU in Russia's stockpile. Plans to further reduce the nuclear arsenals of the United States and Russia are positive.

But we still face immense challenges:

- My long-standing and serious concerns for vastly improved controls over, and reductions in, tactical nuclear weapons remain unresolved.
- The amounts of Russian fissile materials not under adequate control are far too large.
- Promising programs for conversion of Russian weapons scientists to commercial endeavors have been complicated by limited access to the closed cities.
- Far too many radioactive sources are poorly controlled and present significant risks for use in so-called dirty bombs by terrorist groups.
- Too many reactors around the world remain fueled with weapons-grade uranium.
- Too many of our programs focus solely on the Russian Federation, instead of recognizing that we should be working as coleaders with the Russians in a global partnership for nonproliferation programs that encompasses all willing nations.

And of greatest personal disappointment, the absolutely vital plutonium disposition program remains stalled with wrangling over the legal issue of liability indemnification. Frankly, I can only say "Shame on both our nations" when we allow an issue like liability indemnification to stop progress toward disposition of weapons-grade plutonium. Where are our priorities?

This program isn't just in U.S. interests. It isn't just in Russian interests. It is a concrete step toward a safer world. It is incredibly frustrating to watch the endless negotiations on this point while we forget the importance of the underlying program. This begs for leadership from the governments of both our great nations.

In this past year, I became chairman of the Senate Energy and Natural Resources Committee. In that role, I've invested immense time in crafting a national energy policy. As a nation, we face an immense challenge in meeting our growing energy needs while preserving the environment that we cherish.

In the course of this challenge, I've noted how simple my task would be if I only needed to satisfy one region of our nation. But the challenge isn't that simple; we need a policy that meets the needs of the entire country—it can't be done in a piecemeal fashion.

We are poised now to succeed with this new energy policy. After a strong vote in the U.S. House of Representatives, we await a favorable vote in the U.S. Senate. With that vote, the nation will have a comprehensive approach to our diverse needs—an approach that balances conservation, improved efficiencies, and new production.

Achieving increased production, while meeting our environmental mandates, is far from simple. There's no simple or single silver bullet that will solve this dilemma. We must harness all the clean energy sources we have to meet our citizens' needs. Nuclear energy must be a part of that equation.

No other energy source offers the clean, reliable, baseload power that we derive from nuclear energy. Yet for many reasons, development of nuclear energy has been on hold in this nation for over one-quarter of a century. We either reverse that trend now, or we can buy nuclear energy from willing foreign suppliers in not too many years.

For that reason, a significant part of the comprehensive energy bill is devoted to civilian nuclear power. The true words of President Eisenhower echo over the years to remind us of the importance of nuclear power: "The United States knows that peaceful power from atomic energy is no dream of the future. That capability, already proved, is here now—today. Who can doubt . . . that this capability would rapidly be transformed into universal, efficient and economic usage."

In the energy bill, I set out to chart a course over the next decades toward a strong role for nuclear power, both here and around the world, both in developed and in developing nations. I can best summarize my efforts here with discussion of five key goals in the bill.

Of first priority, I had to assure that the liability foundation for our nuclear programs since 1957 remains intact. For that reason, the bill incorporates a twenty-year extension of the Price–Anderson statute.

Then I had to reverse that quarter century dearth of new plant construction. We simply must see new domestic plants constructed. It does us little good to compliment ourselves on our foresight in developing and licensing advanced reactor designs—designs which are providing reliable power around the world today—when we have none of these reactor designs here.

Thus my second priority is to break this mold and provide production tax credit incentives for construction of a few modern plants.

With those few plants, we can show the public that nuclear power plants can be built in this country, and that they can be built with economics and safety consistent with the public's demands. With those few plants, we can convince the investment community that new plant construction is a solid investment opportunity, not one to be shunned.

My third and fourth priorities are essentially tied in importance. They are to reestablish our technology base to meet new needs with new reactors, and develop better waste management solutions.

The bill authorizes a major new R&D reactor program to demonstrate a new generation of ultrasafe, ultraefficient reactors, which minimize waste production and proliferation concerns. It further demands that the project enable R&D for advanced approaches to both electricity and hydrogen production.

As an aside, let me note that progress toward a hydrogen economy is another major emphasis in the energy bill. The opportunity to convert our transportation system to hydrogen fuel offers immense benefits, but only if we can produce that hydrogen cleanly and economically.

As you all know, we can't mine or drill for hydrogen—we have to produce it. That requires energy. The bill supports alternative production approaches, from solar to nuclear to biological.

In the area of nuclear waste policy, I'm convinced that our citizens demand better solutions to nuclear waste than we have today. Now we plan to simply bury and forget about our spent fuel—never mind that it retains an immense store of energy or that its constituents are highly toxic.

There are better solutions, and I've championed their study over the last few years. With the energy bill, we authorize expanded programs to develop better strategies for spent fuel management. These strategies will go far beyond Yucca Mountain and will result in better use of, and far less toxicity from, any future repository.

And finally, my fifth priority in the bill is my recognition that we don't have human resources in the pipeline to support an expansion of nuclear energy, much less to appropriately control and mitigate the military challenges of the nuclear genie.

We need stronger university programs, with more students challenged and motivated to master these technologies to contribute to the future workforce. For that reason, there are strong educational programs in the bill.

I'm confident that we'll pass this bill early next year, and use it to work toward fulfillment of more of Ike's vision.

The bill will enable our nation to proceed with development of nuclear power here at home. But my vision, like Ike's, doesn't stop at our own borders. We must help provide developing nations with the energy resources so that they too can grow and prosper. The seeds of unrest and terrorism will be far less fertile as the standards of living of all peoples are raised toward our own.

As we assist these nations, we should suggest that they focus their attentions on clean energy sources to avoid the environmental problems that many developed nations like the United States are experiencing with past energy sources. Nuclear should be one of the clean technologies we offer to them.

Now obviously, many of those developing nations do not have the necessary infrastructure to produce and safeguard nuclear materials or to design new reactors. But we can and should help them in specific ways, such as providing small sealed reactors, along with full assurances from the developed nations to guarantee to provide all their life-cycle fuel services. Those same life-cycle fuel services should be offered to many other nations for large reactors as well.

We should work through international organizations, like the IAEA that Ike created, to craft global approaches to fuel production and waste handling in ways that minimize proliferation concerns. We just require the will to work through and with the international community to make them a reality.

President Eisenhower offered an immense challenge to the world. And while parts of his vision have been realized, many parts of it remain to fulfill.

Dr. Susan Eisenhower emphasized this view in a recent speech when she noted that 'Atoms for Peace' is a vision—not a blueprint."

Today we need to redouble our efforts to realize that vision and fill in the details of that blueprint, which Ike summarized so well when he closed his speech with the words: "[T]he United States pledges before you—and therefore before the world . . . to devote its entire heart and mind to find the way by which the miraculous inventiveness of man shall not be dedicated to his death, but consecrated to his life."

Appendix C

Some Key Nonproliferation Events from the 1990s to the Present

December 12, 1991	Soviet Nuclear Threat Reduction Act of 1991: Nunn–Lugar Act becomes law
December 25, 1991	Gorbachev resigns signaling breakup of the Soviet Union
October 1992	Senate takes lead in appropriating $400 million for nonproliferation activities
August 1992	President Bush initials the agreement to buy Russian HEU
October 1992	EPACT creates USEC and provides a legal framework for the United States to purchase Russian HEU
February 18, 1993	Presidents Clinton and Yeltsin sign the HEU Agreement for U.S. purchase of 500 MT HEU over a twenty-year period
July 1, 1993	USEC assumes DOE's uranium enrichment executive functions
1994	LEU delivery contract terms established
May 1995	Senator Domenici sponsors legislation to allow forward sales in the United States of the uranium content of Russian HEU
June 1995	First deliveries of LEU made in the United States
Summer 1995	Senator Domenici and the administration work out HEU and USEC privatization compromise
September 1995	Senate passes Domenici's forward sales legislation
November 1995	USEC makes its first shipment of Russian LEU to a customer
April 26, 1996	Senator Domenici's USEC Privatization Act passes

August 1996	USEC turns down Russia's offer to increase the levels of LEU deliveries to 18 MT from the current annual rate of 12 MT
1996	Nunn–Lugar–Domenici Act passes expanding cooperative threat reduction activities
1996	Nuclear Cities program begins
October 1997	Senator Domenici delivers the speech, "A New Nuclear Paradigm" at Harvard
February 1998	Senator Domenici writes to Vice President Al Gore questioning USEC's interests in profits as opposed to nonproliferation goals
March 13, 1998	Senator Domenici writes to Adamov regarding the acceleration of the dismantlement of nuclear weapons
June 26, 1998	Senator Domenici writes to Sandy Berger regarding his concerns about USEC's uranium holdings and the U.S. government's commitment to nonproliferation goals
June 27–July 5, 1998	Senator Domenici travels to France, Russia, and Germany to discuss Eurofab and Euroburn
July 4, 1998	Senator Domenici delivers keynote speech "Youth and the Global Political Challenge of Plutonium" at Obninsk
July 17, 1998	Adamov writes to Senator Domenici regarding Russia's concerns with the HEU agreement due to the privatization of USEC
July 27, 1998	Senator Domenici meets with President Clinton to discuss his proposals for Russian plutonium disposition
July 28, 1998	USEC becomes a private sector entity
September 1–2, 1998	Senator Domenici attends the Moscow Summit with President Clinton and Yeltsin announces his intentions to pull out of the HEU Agreement
September 8, 1998	Senator Domenici writes to President Clinton proposing a compromise in order to save the HEU Agreement from collapsing
October 19, 1998	Congress passes an appropriation saving the HEU Agreement
October 26, 1998	President Clinton signs the appropriations legislation
January 2001	Baker–Cutler Task Force warns the United States faces catastrophic terrorism threat

September 11, 2001	United States attacked by terrorists in New York, Washington, Pennsylvania
November 13, 2001	Bush–Putin Summit announces reductions in nuclear arsenals
Fall 2001	National Security Council report on U.S. nonproliferation programs issues a critical appraisal of plutonium disposition programs in Russia
November 13, 2001	Senator Domenici and other senators send letter to President Bush showing bipartisan support for Russian program activities
January 9, 2002	President Bush promises Senator Domenici support and increased funding for the Russian nonproliferation programs
December 2, 2002	Domenici–Biden–Lugar Nonproliferation Act signed into law as an amendment to the Defense Authorization Bill
June 2003	Evian Summit with the G-8 agreement endorsing Russia's plutonium disposition program "Global Partnership Initiative against the Spread of Weapons and Materials of Mass Destruction"
Beginning of twenty-first century	Senator Domenici spearheaded significant funding increases for nonproliferation programs with Russia: in fiscal 2002, $458.5 million; in fiscal 2003, $812.5 million; in fiscal 2004, $1.28 billion
November 4, 2003	Senators Domenici, Lugar, and Rumyantsev discuss plutonium disposition and try to break the liability issue deadlock
May 17, 2004	Senators Domenici, Feinstein, Lugar, Biden, Alexander, Bingaman and Reed propose amendment to accelerate removal or security of nuclear materials worldwide

Notes

Chapter 1

1. President Dwight D. Eisenhower, "Atoms for Peace." Address to the 470th Plenary Meeting of the United Nations General Assembly (December 8, 1953, New York).

2. March 30, 1953: the Mark I reactor reached criticality, marking the first production of significant quantities of useful nuclear power in the world. *The United States Naval Nuclear Propulsion Program*, March 2003, Department of Energy and the Department of the Navy, 43.

3. Richard Rhodes and Denis Beller, "The Need for Nuclear Power," *Foreign Affairs* 79, no. 1 (January/February 2000).

4. International Energy Agency, *World Energy Outlook 2002*, chapter 13, "Energy and Poverty," IEA, September 2002, 6.

5. Countries possessing nuclear fuel cycle capabilities include France, Russia, Japan, the United Kingdom, China, Canada, Australia, Germany, North Korea, Iran, India, Pakistan, South Africa, the Netherlands, Israel, South Korea, Brazil, and Argentina.

Chapter 2

1. *Congressional Record*, March 30, 1988, 100th Cong., vol. 134, no. 42.

Chapter 3

1. *Global Primary Energy Use*, figure 4, World Energy Council. Online: www.worldenergy.org/wec-geis/downloads1annex_2(1).pdf. For comparison, in the past thirty years energy consumption almost doubled from 207 to 404 quads, while population increased by 62 percent, from 3.7 billion to over 6 billion.

2. U.S. Census Bureau International Data Base, 2003.

3. G. C. Campbell, "The Imminent Peak of World Oil Production" presentation to the U.K. House of Commons, July 7, 1999; updated December 2000 at the Technical University of Claustthal. Online: www.hubbertpeak.com/campbell/commons.html .

4. The joule is the standard international unit of work. One joule is the amount of work done when an applied force of one Newton moves through a distance of one meter in the direction of the force. The Newton is the international system (SI system) unit of force. One Newton is the force required to give a mass of 1 kilogram an acceleration of 1 meter per second. It is named after the English mathematician and physicist Sir Isaac Newton (1642–1727).

5. The WEC is a nonprofit organization created in 1924 to bring sustainable supplies of energy to the world's population through a program of education, conferences, and analytical reports. It is comprised of approximately ninety member committees, geographically defined by country, whose members include energy experts from government entities, commercial sector, and nongovernmental organizations.

6. Department of Energy/Energy Information Administration, *Annual Energy Outlook 2004 with Projections to 2025*, DOE/EIA-0383 (2004), January 2004, appendix A, table A2. Online: www.eia.gov/oiaf/aeol/forecast.html.

7. Department of Energy/Energy Information Administration, *International Energy Outlook 2004*, DOE/EIA-0484 (2004), table A1.

8. DOE/EIA, *International Energy Outlook 2004*, "Highlights," 1.

9. "Global Energy Scenarios to 2050 and Beyond," table 1, World Energy Council. Online: www.worldenergy.org/wec-geis/edc/scenario.asp [accessed June 2004].

10. DOE/EIA, *Annual Energy Outlook 2004*, table A1.

11. DOE/EIA, *Annual Energy Outlook 2004*.

12. "World Energy Needs and Nuclear Power," Nuclear Issues Briefing Paper 11, Uranium Information Centre Ltd., July 2002. Online: www.uic.com.au/nip11.htm [accessed June 2004].

13. DOE/EIA, *International Energy Outlook 2004*, "Highlights," 4.

14. Massachusetts Institute of Technology, *The Future of Nuclear Power, An Interdisciplinary MIT Study*, MIT, 2003. Online: www.web.mit.edu/nuclearpower/ [accessed June 2004]. See figure A-2.2 for 39 billion kWh per year projection and tables A-2.1a–e for current consumption figures.

15. MIT, *The Future of Nuclear Power*. An increase of 13,000 to 39,000 in fifty years is 2.1 percent.

16. John M. Ryskamp, "The Need for Nuclear Power," Idaho National Engineering and Environmental Laboratory, IEEE Power Engineering Society Meeting, April 28, 2003, slide 2.

17. DOE/EIA, *International Energy Outlook 2004*, "Highlights," 4, and MIT, *The Future of Nuclear Power*, respectively.

18. DOE/EIA, *Annual Energy Outlook 2004*, table A8.

19. DOE/EIA, *Annual Energy Outlook 2004*, table A9.

20. Ryskamp, "The Need for Nuclear Power."

21. DOE/EIA, *International Energy Outlook 2004*, "Highlights," 3.

22. DOE/EIA, *Annual Energy Outlook 2004*, "Highlights," 2.

23. DOE/EIA, *Annual Energy Outlook 2004*, DOE/EIA, table A2.

24. DOE/EIA, *Annual Energy Outlook 2004*, table A11.

25. DOE/EIA, *Annual Energy Outlook 2004*, 7.

26. "Shell Cuts Proved Reserves by 20 Percent; Move Sends Shock Waves through Oil Industry," *Petroleum News*, January 9, 2004.

27. AP Wire Service report, March 26, 2004.

28. See www.worldenergy.org/wec-geis/publications/reports/ser/overview.asp.

29. C. J. Campbell, "The Imminent Peak of World Oil Production."

30. DOE/EIA, *International Energy Outlook 2004*, "Highlights," 4.

31. DOE/EIA, *International Energy Outlook 2004*, table A5.

32. *The Global Liquefied Natural Gas Market: Status & Outlook*, DOE/EIA-0637 (2003), December 2003. Online: www.eia.doe.gov/oiaf/analysespaper/global/.

33. "Statement of Alan Greenspan," Chairman, Board of Governors of the Federal Reserve System before the Joint Economic Committee, May 21, 2003, 3–4 (emphasis added).

34. DOE/EIA, *International Energy Outlook 2004*, table A2.

35. DOE/EIA, *International Energy Outlook 2003*.

36. "After Long Taking Its Lumps, Coal Is Suddenly Hot Again," *Wall Street Journal*, April 1, 2004, A1.

37. "America's New Coal Rush," *Christian Science Monitor*, February 26, 2004.

38. See www.aeci.org/generation/newmadrid/nmadrid.html [accessed June 2004].

39. See www.aeci.org/generation/newmadrid/nmadrid.html [accessed June 2004]; estimate based on the emission of 1,474 pounds of carbon per ton of electric utility coal.

40. "Emissions of Criteria Pollutants, CO_2 and Mercury—2000," Missouri Department of Natural Resources. Online: www.aeco.org/generation/newmadrid/nmadrid/html. [accessed June 2004].

41. National Council on Radiation Protection and Measurements, report no. 92 (12/87), report no. 93 (9/87), report no. 94 (12/87), and report no. 95 (12/87). Online: www.yankee.com/license_Radiation.html.

42. Thomas J. Feeley III, "Mercury Reduction in Coal-Fired Power Plants: DOE's R&D Program," NETL, ARIPPA Technical Symposium, August 21, 2002.

43. For renewable energy consumption data see DOE/EIA's *2003 International Energy Outlook*, table A8, and for total world consumption see, table A1.

44. DOE/EIA, *International Energy Outlook 2004*, table A8.

45. DOE/EIA, *Annual Energy Outlook 2004*, table A8.

46. American Wind Energy Association, "The Most Frequently Asked Questions about Wind Energy," American Wind Energy Association 2002, 20. Online: www.awea.org.

47. DOE/EIA, *Annual Energy Outlook 2004*, table A17.

48. DOE/EIA, *Annual Energy Outlook 2004*, table A8.

49. DOE/EIA, *Annual Energy Outlook 2004*, table A17.

50. See www.canwea.ca [accessed June 2004].

51. Robert L. Bradley Jr., "Renewable Energy: Not Cheap, Not Green," *Cato Policy Analysis* 28 (August 27, 1997). Online: www.cato.org/pubs/pas/pa-280.html.

52. Renee Mickelburgh, Tony Paterson, and Kim Willsher, "Wind Farms Feel the Chill of Public Rejection," April 5, 2004. Online: www.smh.com.au.artivles/2004/04/04/1081017039062.html [accessed April 2004].

53. "Island Turbines Whip up a Storm," January 2, 2004. Online: www.production1.energycentral.com/centers/news/daily [accessed January 2004].

54. Mickelburgh, Paterson, and Willsher, "Wind Farms Feel the Chill of Public Rejection."

55. "When Blade Meets Bat," *Scientific American*, February 2004, 20–21.

56. DOE/EIA, *International Energy Outlook 2003*, 115.

57. American Wind Energy Association, "The Most Frequently Asked Questions about Wind Energy."

Chapter 4

1. Richard W. Dyke and Francis X. Gannon, *Chet Holifield: Master Legislator and Nuclear Statesman* (Lanham, Md.: University Press of America, 1996), 16.

2. World Nuclear Association, "World Nuclear Power Reactors 2002–2004 and Uranium Requirements," World Nuclear Association, Information and Issue Briefs, March 25, 2004. Online: www.world-nuclear.org/info/reactors.htm [accessed June 2004].

3. N. Nakicenovic, A. Gruben, and A. McDonald, eds., *Global Energy Perspectives* (London: Cambridge University Press, 1998).

4. MIT, *The Future of Nuclear Power*, 3.

5. International Atomic Energy Agency, *Energy, Electricity and Nuclear Power Estimates for the Period up to 2030* (Vienna: IAEA, July 2003), table 3.

6. DOE/EIA, *International Energy Outlook 2003*, table 20, 102.

7. DOE/EIA, *International Energy Outlook 2004*, table E1.

8. DOE/EIA, *International Energy Outlook 2004*, tables E2 and E3, respectively.

9. "Industry Briefs Wall Street on Plant Efficiency, Growth Plans," *Nuclear Energy Insight*, March 2004.

10. *Nucleonics Week*, April 8, 2004.

11. DOE/EIA, *Annual Energy Outlook 2004*, table A8.

12. DOE/EIA, *Annual Energy Outlook 2004*, table A9.

13. David W. South, "Treatment of Nuclear Power in EIA/NEMS: Current Representation and Recommended Changes" (briefing for Energy Information Administration, December 2, 1999, made available by Energy Resources International on June 22, 2004).

14. For example, for the 1999 Annual Energy Outlook, the EIA assumed 1995–1997 averaged operating cost data, $150 per kW cost for steam generator replacement in the thirtieth year of plant operation and a $250 per kW cost in the fortieth plant year for equipment replacement for those plants it assumed would seek license renewal. The EIA assumed license renewal costs were more than twice as high as those estimated by the NRC at the time (NUREG-1437).

15. DOE/EIA, *Annual Energy Outlook 2004*.

16. Hermann A. Grunder, B. D. Shipp, Michael R. Anastasio, Pete Nanos, William J. Madia, and C. Paul Robinson, *Nuclear Energy: Power for the Twenty-first Century*, Argonne National Laboratory report no. ANL-03/12, Sandia National Laboratory Report no. SAND3002-1545P, May 2003. Online: www.nuclear.inel.gov/papers-presentations/power_for_the_21st_century.pdf [accessed June 2004].

17. Office of Nuclear Energy, Science, and Technology, "FY 2004-2008 Budget Strategy" (presentation to the undersecretary for energy, science, and environment, Office of Nuclear Energy, Science and Technology, April 18, 2002).

18. See www.nei.org/doc.asp?docid=1184 [accessed June 2004].

19. EnergyWashington.com daily updates, May 29, 2003.

20. Press release, Svensk Karnbranslehantering ABSKB, July 11, 2003.

21. Alain Bugat, "Atoms for Peace +50: Nuclear Energy and Science for the Twenty-first Century" (speech delivered to the French Atomic Energy Commission, October 22, 2003).

22. Canadian Nuclear Association, "Nuclear Canada," *Canadian Nuclear Association Electronic Newsletter* V, no. 12 (March 26, 2004): 1. Online: www.can.ca.

23. Canadian Nuclear Association, "Nuclear Canada."

24. Planet Ark, online: Planetark.org [accessed June 2004].

25. *Business Standard*, November 13, 2003. Online: www.business-standard.com.

Chapter 5

1. Richard Wilson, "The Changing Need for a Breeder Reactor," Uranium Institute Annual Symposium, London, September 1999.

2. Senate Energy and Water Development Appropriation Report, 105th Cong., 2nd sess., 1999, S. Rep. 105-206 (June 5, 1998), 148.

3. NRC generic letters are transmittals to request an action, circulate technical or policy positions, or solicit participation in voluntary programs.

4. Center for Strategic and International Studies, *Executive Summary, The Regulatory Process for Nuclear Power Reactors: A Review*, Center for Strategic and International Studies (Washington, D.C., 1999), 10.

5. Shirley Jackson, chairman, Nuclear Regulatory Commission, to the Subcommittee on Clean Air, Wetlands, Private Property and Nuclear Safety of the Committee on Environment and Public Works on July 30, 1998, 5. Online: epw.senate.gov/105th/nrca-7-3.htm [accessed January 2004].

6. Shirley Jackson, statement given at the Twenty-sixth Annual Water Reactor Safety Meeting, October 26, 1998. Online: www.nrc.gov/reading-rm/doc-collections/commission/speeches/1998/s98-26.html [accessed January 2004].

7. NRC fact sheets. Online: www.nrc.gov/reading-rm/doc-collections/fact-sheets/ [accessed June 2004].

8. Testimony on *Radiation Standards: Scientific Basis Inconclusive, and EPA and NRC Disagreement Continues*, GAO/T-RECD-00-252, before the Subcommittee on Science, U.S. House of Representatives, July 18, 2000.

9. English translation, Andre Aurengo, on behalf of a working group, National Academy of Medicine, Paris, France (emphasis added). Online: www.cnts.wpi.edu/RSH/French_Academy_of_Medicine-2003.htm.

10. Bernard L. Cohen, "Validity of the Linear No-Threshold Theory of Radiation Carcinogen Is at Low Doses" (speech given at the Uranium Institute's Twenty-third Annual Symposium, September 10–11, 1998, London, England).

11. Kenneth L. Mosman et al., "Bridging Radiation Policy and Science, An International Conference, Final Report" (speech delivered in Warrenton, VA, January 31, 2000). Online: www.cnts.wpi.edu/RSH/DOCS/BRPS%2520FINAL.pdf.

12. Environmental Protection Agency, *Estimating Radiogenic Cancer Risks*, EPA 402-R-93-076, EPA, June 1994.

13. A rem is a radiation dose expressed in terms of its biological effects. Background radiation per year for an average American is estimated to be in the range of 300 to 400 millirems per year. A millirem is 1/1,000th of a rem.

Chapter 6

1. DOE/EIA, *Uranium Industry Annual 1991*, DOE/EIA-0478(91), October 1992, 36.
2. *Statistical Data of the Uranium Industry*, GJO-100 (77), 91.
3. *Statistical Data of the Uranium Industry*, GJO-100 (83), 41–45.
4. *Statistical Data of the Uranium Industry*, GJO-100 (83), 41.
5. *Statistical Data of the Uranium Industry*, GJO-100 (83), 67.
6. The "Redbook" is *Uranium 2001: Resources, Production and Demand* (a joint report by the OECD Nuclear Energy Agency and the International Atomic Energy Agency), 2002.
7. The International Generation IV Forum comprises Argentina, Brazil, Canada, the European Union, France, Japan, the Republic of Korea, the Republic of South Africa, Switzerland, the United Kingdom, and the United States.

Chapter 7

1. Nuclear Energy Research Advisory Committee (NERAC), Office of Nuclear Energy, Science and Technology, U.S. Department of Energy, March 2003. Online: www.nuclear.gov [accessed June 2004].
2. Michael L. Corradini et al., "The Future of University Nuclear Engineering Programs and University Research and Training Reactors" (report presented to the Nuclear Energy Research Advisory Committee, May 10, 2000).
3. James F. Stubbins, statement for the Hearing on University Resources for the Future of Nuclear Science and Engineering Programs before the Energy Subcommittee, House Committee on Science, 108th Cong., 1st sess. (June 10, 2003), 3. Online: www.house.gove/science/hearings/energy03/index.htm [accessed September 2003].
4. David M. Slaughter, "Developing New Paradigms to Improve Educational Experiences and Support Unique Infrastructure in Nuclear Engineering and Nuclear-Related Disciplines," Hearing on University Resources for the Future of Nuclear Science and Engineering Programs before the Energy Subcommittee, House Committee on Science, 108th Cong., 1st sess. (June 10, 2003), 1. Online: www.house.gove/science/hearings/energy03/index.htm [accessed September 2003].
5. The U.S. system of higher education is not alone in this predicament—top German and French nuclear engineering schools have experienced a loss of nuclear curriculum, professors, and declining enrollments. *Nucleonics Week*, October 16, 2003, 15–16.
6. Angelina S. Howard, "Testimony before the Energy Subcommittee, House Committee on Science," Hearing on University Resources for the Future of Nuclear Science and Engineering Programs before the Energy Subcommittee, House Committee on Science, 108th Cong., 1st sess. (June 10, 2003), 10. Online: www.house.gove/science/hearings/energy03/index.htm [accessed September 2003].
7. J. P. Freidberg, "Nuclear Engineering in Transition: A Vision of the Twenty-first Century," *Nuclear News* (June 1999): 50.
8. Corradini et al., "The Future of University Nuclear Engineering Programs," 9.
9. Corradini et al., "The Future of University Nuclear Engineering Programs," 8.
10. Daniel M. Kammen, "The Future of University Nuclear Science and Engineering Programs," testimony for the Hearing on University Resources for the Future of

Nuclear Science and Engineering Programs before the Energy Subcommittee, House Committee on Science, 108th Cong., 1st sess. (June 10, 2003), 12. Online: www.house .gove/science/hearings/energy03/index.htm [accessed September 2003].

11. Slaughter, "Developing New Paradigms to Improve Educational Experiences," 1.

12. Stubbins, statement for the Hearing on University Resources for the Future of Nuclear Science and Engineering Programs.

13. Nuclear Energy Institute, *Nuclear Pipeline Analysis*, NEI, December 2001.

14. 2003 Nuclear Industry Staffing Survey, private communication from the Nuclear Energy Institute, April 2004.

15. "Utility Execs Peer into the Crystal Ball," *Nuclear News*, April 2003, 55.

16. "Utility Execs Peer into the Crystal Ball."

17. 2003 Nuclear Industry Staffing Survey, private communication from the Nuclear Energy Institute, April 2004.

18. Greta Joy Dicus, "The Changing Nuclear Workforce" (speech delivered to the Oak Ridge Women in Nuclear Chapter, Oak Ridge, Tennessee, November 30, 2000). Online: www.nrc.gov/reading-rm/doc-collections/commission/speeches/2000/s00-30. html [accessed August 2003].

19. Kammen, "The Future of University Nuclear Science and Engineering Programs."

20. Grunder et al., *Nuclear Energy: Power for the Twenty-first Century*.

21. Letter from senators to Honorable Federico Peña, U.S. Department of Energy, July 30, 1997.

22. Secretary of Energy Federico Peña responded to our letter on September 16, 1997 agreeing with our position saying, "During the next several weeks . . . Dr. Terry Lash . . . will hold discussions with our National Laboratories, U.S. universities, and industry to establish the basis of a new nuclear energy program." In November 1997, the President's Committee of Advisors on Science and Technology report on the "Federal Energy Research and Development for the Challenges of the Twenty-first Century" recommended that the Department of Energy initiate the Nuclear Energy Research Initiative. It is a jointly funded DOE-industry program.

23. Proposed "Energy Policy Act," S.2095, February 12, 2004

24. Stubbins, statement for the Hearing on University Resources for the Future of Nuclear Science and Engineering Programs.

Chapter 8

1. Nuclear proliferation refers to the acquisition or production of materials (either highly enriched uranium or plutonium) for the purpose of constructing a nuclear weapon, the acquisition of nuclear materials (which may or may not be weapons-useable) for use in a radiological bomb, or the acquisition of knowledge related to weapons production.

2. Senator Pete Domenici, "A New Nuclear Paradigm" (speech before the Inaugural Symposium Belfer Center for Science and International Affairs, Harvard University, October 31, 1997); see appendix A.

3. North Korea obtained high-grade plutonium from a production reactor, not a civilian LWR. Gas-cooled or heavy water moderated reactors could be sources of weapons quality plutonium. However, no country has pursued this route since India purportedly did.

4. This plant has encountered some operating problems.
5. George W. Bush, "Address on Weapons of Mass Destruction Proliferation" (remarks by the president at the National Defense University, Fort Lesley J. McNair, Washington, D.C., February 11, 2004). Online: www.whitehouse.gov/news/releases/2004/02/print/20040211/-4/html [accessed February 2004].
6. "Conference Report Making Appropriations for Energy and Water Development for the Fiscal Year Ending September 30, 2004, and for Other Purposes," 108rh Cong.,1st sess., S. Rep. 108-357, 163.
7. Congressional Record 137, S.16486-01.
8. Howard Baker and Lloyd Cutler, "Final Report, Task Force on DOE Nonproliferation Programs in Russia," January 2001. Online: www.eisenhowerinstitute.org/programs/globalpartnerships/safeguarding/threatreduction [accessed August 2003].
9. David E. Sanger, "Pakistani Says He Saw North Korean Nuclear Devices," New York Times, April 13, 2004, A12.
10. "Progress Report US-Russian Megatons to Megawatts Program Recycling Nuclear Warheads into Electricity as of December 31, 2003." Online: www.usec.com/v2001_02/HTMLmegatons-status.asp [accessed June 2004].
11. "The Nunn–Lugar Vision, 1992–2002," Nuclear Threat Initiative, Washington, D.C. Online: www.nti.org [accessed June 2004].
12. World Nuclear Association, "Research Reactors," Information and Issue Briefs, World Nuclear Association, August 2003.
13. President Bush, "Address on Weapons of Mass Destruction Proliferation."
14. General Dr. Mohamed El Baradei, "Towards a Safer World," Economist, October 2003, 43–44.

Chapter 9

1. Raymond LeRoy Murphy, Understanding Radioactive Waste, 3rd ed., ed. Judith A. Powell (Columbus, Ohio: Batelle Press, 1989), 57.
2. Murphy, Understanding Radioactive Waste, 58.
3. Murphy, Understanding Radioactive Waste, 58.
4. The League of Women Voters, The Nuclear Waste Primer: A Handbook for Citizens (New York: Lyons & Burford, 1993), 19, 24.
5. Information provided by Dr. Louis Rosen, Los Alamos Laboratory Fellow and retired Director, Los Alamos Meson Physics Facility, June 24, 2003.
6. "Protection against Radioactivity Is Easy," slide from Areva Technical Days, June 27–28, 2002, Paris, France.
7. Private communication from Cogema on November 24, 2003. Each glass canister contains on average 84 kg of fission products and actinides and is equivalent to 360 million kWh or electrical power consumption.
8. Information provided by Dr. Louis Rosen, Los Alamos Laboratory Fellow and retired director, Los Alamos Meson Physics Facility, June 24, 2003.
9. Private communication from Cogema on November 24, 2003.
10. Utilities have filed over twenty-five lawsuits, mostly in the Federal Claims Court seeking almost $50 billion of damages due to the DOE's refusal to start accepting spent nuclear fuel for ultimate disposal by January 31, 1998. The U.S. Court of Appeals has established the government's liability for the D.C. Circuit and the Federal

Circuit. The utilities and the DOE continue to argue over the DOE's obligation to pay for storage costs in the wake of the DOE not picking up the spent fuel, a schedule to start taking the fuel, and damages the DOE should pay the utilities.

11. "Radiation Standards, Scientific Basis Inconclusive, and EPA and NRC Disagreement," Washington, D.C., General Accounting Office, GAO/T-RCED-00-252, July 2000.

12. "Report and Recommendations of the Nevada Commission on Nuclear Projects" (presented to the governor and legislature of the State of Nevada, December 2002), 18. Online: www.state.nv.us/nucwaste/news2001/commrpt2000.pdf.

13. At the current stage of the process the licensing issues include: how fast water might travel through Yucca Mountain and seep into the tunnels where nuclear waste canisters will be stored; the chemical environment within the drifts that might cause the containers to corrode and allow the radioactive materials to escape into the environment; and the potential impact of earthquake or volcanic activity near the site.

14. U.S. Department of Energy Office of Nuclear Energy, Science, and Technology, "Report to Congress on Advanced Fuel Cycle Initiative: The Future Path for Advanced Spent Fuel Treatment and Transmutation Research," January 2003, I-5. Online: www.ne.doe.gov/reports/AFCI_CongRpt2003.pdf.

15. U.S. Department of Energy, "Report to Congress on Advanced Fuel Cycle Initiative," I-3. There are 44,000 MT of spent nuclear fuel that today contain 440 MT of plutonium.

16. Private communication from Cogema on November 24, 2003.

17. Rhodes and Beller, "The Need for Nuclear Power," 38.

18. Actinides, elements in the periodic table with atomic numbers between 89 and 104, are created by the capture of neutrons. They are often referred to as transuranics, all of those elements with atomic numbers higher than uranium. See U.S. Department of Energy, "Report to Congress on Advanced Fuel Cycle Initiative."

19. U.S. Department of Energy, "Report to Congress on Advanced Fuel Cycle Initiative."

20. Burton Richter, "A Personal Commentary on the MIT Nuclear Power Study," private communication, March 15, 2004.

21. Richter, "A Personal Commentary on the MIT Nuclear Power Study."

22. Argonne National Laboratory Reactor Analysis and Engineering Division, Online: www.rae.anl.gov/research/afc/system_studies [accessed June 2004].

23. National Energy Policy Development Group, "Reliable, Affordable, and Environmentally Sound Energy for America's Future," May 17, 2001. Also referred to as the Bush National Energy Policy.

24. U.S. Department of Energy, "Report to Congress on Advanced Fuel Cycle Initiative," III-4.

25. Office of Nuclear Energy Science and Technology FY 2004 Congressional Budget, 45. Online.

26. Richter, "A Personal Commentary on the MIT Nuclear Power Study."

27. "Business Case for New Nuclear Power Plants," *Scully Capital*, final draft, July 2002, 6-4.

28. U.S. Department of Energy, "Report to Congress on Advanced Fuel Cycle Initiative," E-3.

29. Cogema's critique of the MIT study reports that the incremental cost of MOX recycling is between 4 percent and 6 percent of the kWh cost, depending on whether the base case (6.7 cents/MWh) or the "most optimistic plausible case" (4.2 cents/MWh) is considered. Dr. Burton Richter states that a recent report by the Harvard Project on Managing the Atom gives the cost penalty for using MOX compared to fresh fuel as 0.13 cents per kWh in the price of electricity. He calls this "noise" compared to the other sources of potential price volatility!

Chapter 10

1. Grunder et al., *Nuclear Energy: Power for the Twenty-first Century*.
2. Alan D. Pasternak, "Global Energy Futures and Human Development: A Framework for Analysis," UCRL-ID-140773, U.S. Department of Energy, Lawrence Livermore National Laboratory, October 2000.
3. IEA, *World Energy Outlook 2002*, chapter 13 (5).
4. Sir Bernard Ingham, "Nuclear's Presentational Problem" (speech presented to the World Nuclear Association Annual Symposium 2002). Online: www.world-nuclear.org/sym/2002/ingham.htm [accessed March 2004].
5. "Towards a European Strategy for the Security of Energy Supply," European Commission, November 2002. Online: www.europe.eu.int/comm/energy_transport/en/lip_en.html [accessed August 2003].
6. Ingham, "Nuclear's Presentational Problem."
7. Atlantic Council of the United States, "Clean Air for Asia: China-India-Japan-United States Cooperation to Reduce Air Pollution in China and India," Atlantic Council of the United States, Washington, D.C., July 2003, viii.
8. *Atlantic Council of the United States "Clean Air for Asia,"* 4–5.
9. DOE/EIA, *Annual Energy Outlook 2004*, table A2.
10. DOE/EIA, *Annual Energy Outlook 2004*, table A1.
11. DOE/EIA, *Annual Energy Outlook 2004*, table A1.
12. Nuclear Energy Institute, "Why a Vision for Nuclear Energy," NEI. Online: www.nei.org.
13. The conference agreement regarding medical isotopes infrastructure includes $28,425,000 for the medical isotope program. From within available funds, the department is directed to provide $4,000,000 for upgrades of radiological facilities at Oak Ridge National Laboratory. See "Energy and Water Appropriations for FY 2004, 108th Cong., 1st sess., H. Rep. 108-357.
14. Colin Hunt, "Comprehensive Assessment of Energy Systems: Severe Accidents in the Energy Sector," executive summary, Paul Scherrer Institute, April 23, 2003, table 9.1.5.
15. Rhodes and Beller, "The Need for Nuclear Power."
16. See muller/lbl.gov/teaching/physics10/pages02fall/risk.html [accessed October 2003].
17. See www.hcra.harvard.edu/index.html [accessed June 2004].
18. Data provided by Admiral F. L. Bowman, U.S. Navy Director, Naval Nuclear Propulsion Program, November 19, 2003.
19. Nuclear Energy Institute, "Public Health Risk Low in Unlikely Event of Terrorism at Nuclear Plant," NEI Fact Sheet, March 2003.

20. See www.csis.org/isp/sv/index.htm regarding the Silent Vector exercise.

21. Two U.S. nuclear-powered submarines have been lost at sea in the Atlantic Ocean, apparently as a result of non-nuclear reactor incidents. The Thresher submarine sank on April 10, 1963, 200 miles off the eastern coast of the United States in water 8,500 feet deep. The *Scorpion* submarine sank on May 22, 1968, 400 miles southwest of the Azores in 10,000 feet of water. Repeated seawater and sediment testing has shown no evidence of release of radioactivity from the reactor fuel elements in either of the submarines. In addition, there is no evidence of leakage of plutonium from nuclear weapons that were onboard the *Scorpion* when it sank. See "Environmental Monitoring and Disposal of Radioactive Wastes from U.S. Naval Nuclear-powered Ships and Their Support Facilities," Department of the Navy, Report NT-03-1, March 2003, 6–7.

22. Department of Energy and Department of the Navy, "The United States Naval Nuclear Propulsion Program, Over 127 Million Miles Safely Steamed on Nuclear Power," March 2003.

23. Rhodes and Beller, "The Need for Nuclear Power," 31.

24. Private communication from Dr. Nebojsa Nakicenovic to Julian Steyn on December 14, 2003.

25. Information provided by Dr. Louis Rosen, Los Alamos Laboratory Fellow, and retired director, Los Alamos Meson Physics Facility, on June 24, 2003.

26. Peter Beck and Malcolm Grimston, "Double or Quits? The Global Future of Civil Nuclear Energy," briefing paper, April 2002, the Royal Institute of International Affairs, London, England. Online: www.riia.org.

27. T. U. Marston, "Nuclear Power's Role in Meeting Environmental Requirements: Evaluation of E-EPIC Economic Analyses," Electric Power Research Institute, final report, Palo Alto, California, January 2003.

28. Nuclear Energy Institute, June 2003, based on 2002 data provided by FERC and proprietary sources.

29. Senator Pete Domenici, "Domenici: Energy Diversity and Increased Production of Clean Energy Will Blunt Natural Gas Crisis," news release, July 10, 2003. Online: www.domenici.senate.gov/newscenter [accessed September 2003].

30. "America Must Consider Nuclear, Greenspan Says," Insight Article, Nuclear Energy Institute, July 2003. Online: www.nei.org.

31. Statement of Alan Greenspan before the Joint Economic Committee of Congress, May 21, 2003. Online: www.federalreserve.gov/boarddocs/testimony/2003 [accessed June 2003].

32. Private communication from Cogema on November 24, 2003. Studies reviewed included "The Reference Costs of the Electricity Production in France," Ministry of Industry, 1997; "Projected Cost of Generating Electricity-Update 1998," by the International Energy Agency and the Nuclear Energy Agency; report of the Parliamentary Office for Assessing Scientific and Technological Choices by Messrs. Galley and Bataille; the "Economic Forecast Study of the Nuclear Option," by Messrs. Charpin, Dessus, and Pellat, 2000; report of the Commission for Analysis of Electricity Generation Technologies and Energy Development, 2000; study released (in Finnish) by Dr. Tarjanne, Lappeenranta University, June 2000; and, reported results of the French Ministry of Industry's update of its 1997 assessment, February 2003.

33. "Japan Study on Power Costs," World Nuclear Association Weekly Digest, March 5, 2004.

34. "Nuclear Canada Canadian Nuclear Association Electronic Newsletter, March 16, 2004, V, no. 11. Online: www.cna.ca.

35. Richter, "A Personal Commentary on the MIT Nuclear Power Study."

36. Massachusetts Institute of Technology, "The Future of Nuclear Power, An Interdisciplinary MIT Study," MIT, 2003. Online: www.web.mit.edu/nuclearpower/ [accessed June 2004], table 5.1.

37. According to Westinghouse, the proposed AP 1000 is faster to build than past plants. Its components can be built in a factory and assembled on-site. It is simpler to build with 50 percent fewer valves, 35 percent fewer pumps, 80 percent less piping, and 70 percent less wiring. The construction cost is estimated at $1.3 billion. It would generate 1,117 MWe.

38. Leon Walters, David Wade, and David Lewis, "Transition to a Nuclear/Hydrogen Energy System," World Nuclear Association Symposium 2002. Online: www.world-nuclear.org/sym/2002/walters.htm [accessed March 2004].

39. Grunder et al., Nuclear Power: Power for the Twenty-first Century, 12.

40. Paul M. Grant, "Hydrogen Lifts Off—With a Heavy Load," Nature 424, no. 6945 (July 10, 2003).

41. William Magwood IV, "The Nuclear Hydrogen Initiative," American Nuclear Science Conference, June 2, 2003, San Diego, California.

42. The Idaho National Engineering and Environmental Laboratory and Argonne National Laboratory West will be merged in early 2005 to form the Idaho National Laboratory. It will coordinate U.S. participation in the Generation IV Nuclear Energy Systems Program and the Advanced Fuel Cycle Initiative.

43. Walters et al., "Transition to a Nuclear/Hydrogen Energy System."

44. "Framatome ANP has a $26 Million Budget This Year to Develop an HTGR," Platts Nuclear News Flashes, January 15, 2004, 3. Online: www.platts.com.

45. Walters et al., "Transition to a Nuclear/Hydrogen Energy System," 7.

46. Presentation by Richard Rhodes before the Nuclear Energy Institute's Nuclear Energy Assembly, Chicago, Illinois, May 3–5, 2000.

47. John W. Simpson, Nuclear Power from Underseas to Outer Space (La Grange Park, Ill.: American Nuclear Society, 1995), 100.

48. MIT, "The Future of Nuclear Power," vii.

49. Statement of Gerald Doucet, Secretary General World Energy Council, April 2, 2003, Berlin, Germany.

50. "Life-Cycle Analysis of Power Generation Systems," Central Research Institute of Electric Power Industry, March 1995. Online: www.nei.org/index.asp?catnum=2&catid=260 [accessed June 2004].

51. Energy Information Administration, "Carbon Dioxide Emission Factors for Coal," EIA, Quarterly Coal Report, January–April 1994, DOE/EIA-0121, Washington, D.C., August 1994.

52. Presentation to the undersecretary for energy, science, and environment, Office of Nuclear Energy, Science and Technology, April 18, 2002.

53. Department of Energy Nuclear Energy Research Advisory Group, "Technology Roadmap for Generation IV Nuclear Energy Systems Executive Summary," DOE GIF-001-00, September 23, 2002, 4.

54. "World Energy, Technology and Climate Policy Outlook," European Commission (Luxembourg: Office for Official Publications of the European Commission, May 2003).

55. Richard Rhodes, "Atoms for Peace Anniversary," Nuclear Energy Assembly 2003, 3.

56. Nuclear Energy Institute, "Nuclear Energy Developments," NEI, June 2003.

57. Quoted in Romney B. Duffey and Alistair I. Miller, "From Option to Solution: The Nuclear Contribution to Global Sustainability Using the Newest Innovations," World Nuclear Association Symposium 2002, September 4–6 (italics added). Online: www.world-nuclear.org/sym2002/pdf/duffey.pdf.

58. Admiral F. L. Bowman, U.S. Navy, Director, Naval Nuclear Propulsion Program (speech at the Atoms for Peace +50 Conference, October 22, 2003, Washington, D.C.).

59. Alain Bugat, Chairman, French Atomic Energy Commission (speech at the Atoms for Peace +50 Conference, October 22, 2003, Washington, D.C.).

60. Denis E. Beller, "Atomic Time Machines" (speech delivered at the Eighth Annual Wallace Stegner Center Symposium "The Nuclear West: Legacy and Future"), *Journal of Hard, Resource and Environmental Law* 24 (2004): 41–60.

61. 80,000 MWe × .9 (nuclear capacity factor) divided by .35 (wind mill capacity factor) = 205 GWe.

62. American Wind Energy Association, "The Most Frequently Asked Questions about Wind Energy," 16.

63. James Lovelock, "The Independent." Online: www.argument.independent.co.uk/commentators/story.jsp?story=52340 [accessed June 2004].

Chapter 11

1. Beller, "Atomic Time Machines."

2. Statement of Admiral F. L. Bowman, U.S. Navy Director, Naval Nuclear Propulsion Program, before the House Committee on Science, 108th Cong., 1st sess. (October 29, 2003). Online: www.house.gov/science/hearings/full03/Oct29/bowman.pdf.

3. As mentioned in chapter 1, these countries include France, Russia, Japan, the United Kingdom, China, Canada, Australia, Germany, North Korea, Iran, India, Pakistan, South Africa, the Netherlands, Israel, South Korea, Brazil, and Argentina.

Index

About the Author

Senator Pete V. Domenici was born in 1932 in Albuquerque, New Mexico, the only son of Italian immigrants. After earning an education degree and teaching in public high school, he earned his law degree from the University of Denver and entered private practice in Albuquerque in 1958. After serving as mayor of Albuquerque, he was elected to the U.S. Senate in 1972 as the first Republican in thirty-eight years to serve from that state. With reelection in 2002, Domenici has served longer in the U.S. Senate than any other New Mexican in history. He and his wife Nancy have eight children.